Designing Cereal and Legume Based Foods with Improved Nutritional Properties

Designing Cereal and Legume Based Foods with Improved Nutritional Properties

Editors

Jennifer Ahn-Jarvis
Brittany A. Hazard

MDPI

Basel • Beijing • Wuhan • Barcelona • Belgrade • Novi Sad • Cluj • Manchester

Editors
Jennifer Ahn-Jarvis
Norwich Research Park
Norwich, UK

Brittany A. Hazard
Norwich Research Park
Norwich, UK

Editorial Office
MDPI
St. Alban-Anlage 66
4052 Basel, Switzerland

This is a reprint of articles from the Special Issue published online in the open access journal *Foods* (ISSN 2304-8158) (available at: https://www.mdpi.com/journal/foods/special_issues/Designing_Cereal_Legume_based_Foods_Improved_Nutritional_Properties).

For citation purposes, cite each article independently as indicated on the article page online and as indicated below:

Lastname, A.A.; Lastname, B.B. Article Title. *Journal Name* **Year**, *Volume Number*, Page Range.

ISBN 978-3-0365-9905-2 (Hbk)
ISBN 978-3-0365-9906-9 (PDF)
doi.org/10.3390/books978-3-0365-9906-9

Contents

foods

MDPI

Article

Mining the A.E. Watkins Wheat Landrace Collection for Variation in Starch Digestibility Using a New High-Throughput Assay

Petros Zafeiriou [1], George M. Savva [1], Jennifer H. Ahn-Jarvis [1], Frederick J. Warren [1], Marianna Pasquariello [2], Simon Griffiths [2], David Seung [2] and Brittany A. Hazard [1,*]

1 Quadram Institute Bioscience, Norwich NR4 7UQ, UK
2 John Innes Centre, Norwich NR4 7UH, UK
* Correspondence: brittany.hazard@quadram.ac.uk; Tel.: +44-(0)1603-255000

Abstract: Breeding for less digestible starch in wheat can improve the health impact of bread and other wheat foods. The application of forward genetic approaches has lately opened opportunities for the discovery of new genes that influence the digestibility of starch, without the burden of detrimental effects on yield or on pasta and bread-making quality. In this study we developed a high-throughput in vitro starch digestibility assay (HTA) for use in forward genetic approaches to screen wheat germplasm. The HTA was validated using standard maize and wheat starches. Using the HTA we measured starch digestibility in hydrothermally processed flour samples and found wide variation among 118 wheat landraces from the A. E. Watkins collection and among eight elite UK varieties (23.5 to 39.9% and 31.2 to 43.5% starch digested after 90 min, respectively). We further investigated starch digestibility in fractions of sieved wholemeal flour and purified starch in a subset of the Watkins lines and elite varieties and found that the matrix properties of flour rather than the intrinsic properties of starch granules conferred lower starch digestibility.

Keywords: wheat; landrace; starch digestibility; natural variation; high-throughput; retrograded starch; starch structure; screening tool; pipetting tool; breeding

Citation: Zafeiriou, P.; Savva, G.M.; Ahn-Jarvis, J.H.; Warren, F.J.; Pasquariello, M.; Griffiths, S.; Seung, D.; Hazard, B.A. Mining the A.E. Watkins Wheat Landrace Collection for Variation in Starch Digestibility Using a New High-Throughput Assay. *Foods* **2023**, *12*, 266. https://doi.org/10.3390/foods12020266

Academic Editor: Wenhao Li

Received: 16 November 2022
Revised: 20 December 2022
Accepted: 22 December 2022
Published: 6 January 2023

1. Introduction

Starchy foods are a main source of carbohydrate in our diet and an important source of energy. Reducing the rate and extent of starch digestibility in foods can help to maintain healthy blood glucose levels, which is important for the prevention and management of obesity and chronic diseases like type II diabetes [1]. Furthermore, there is substantial evidence that consumption of resistant starch, that is the starch that escapes digestion in the small intestine and reaches the colon, can reduce blood glucose and help to maintain a healthy gut [2–4]. Wheat (*Triticum aestivum* L.) is one of the most widely consumed crops worldwide and provides up to 50% of the calories in the human diet, mainly in the form of starch, so improving its nutritional quality could deliver health benefits to a large number of people [5,6]. Thus, breeding for wheat with less digestible starch is an important strategy to develop healthier foods to reduce dietary risk factors of chronic diseases.

A mature wheat grain contains predominately starch, 60–70% (*w*/*w*), which is the greatest contributor to calories, and also 10–15% (*w*/*w*) protein and 11–15% (*w*/*w*) dietary fiber [7,8]. Starch is produced in the endosperm of wheat during grain filling and is composed of two distinct α-glucan polymers, amylose (linear α-1,4-linked chains) and amylopectin (α-1,4-linked chains with α-1,6-linked branches). In their native state, starch polymers exist as partially crystalline granules, which are difficult for humans to digest; thus starch-based foods are typically cooked prior to consumption. Heating starch in the presence of water leads to starch gelatinization making starch polymers more digestible [9–11]. Subsequent cooling leads to retrogradation in which starch polymers form crystalline structures that are less digestible [12,13]. Moreover, there are several other factors which contribute to differences in starch digestibility such as starch molecular structure [14–16] and properties of wheat flour (particle size and protein content) [17,18].

To date, reverse genetic studies in wheat have demonstrated potential for increasing resistant starch levels using induced mutations in starch biosynthesis genes [14,19–21]. However, initial analyses of some mutants have shown detrimental effects on yield and on pasta and bread-making quality, so identifying additional sources of genetic variation for starch digestibility could support the development of improved traits for commercial breeding applications [14,20]. Wheat landraces, locally adapted lines that have not been modified through modern breeding techniques, present reservoirs of genetic diversity that can be introduced into modern varieties. Of special note is the A.E. Watkins bread wheat landrace collection, encompassing 826 bread wheat landraces collected in the 1920s and 1930s from a global geographic distribution. This collection showed a greater level of genetic diversity compared to modern elite European bread wheat varieties and has been used to identify resistance genes for a variety of diseases [22–26]. The Watkins lines have been purified by single-seed descent from which many genomic and genetic resources have been developed. A core set of 118 accessions (c.Watkins) capturing most of the genetic diversity in the Watkins collection was presented by Wingen et al. [22], and used to generate nested association mapping populations, all of which were genotyped, have genetic maps available, and are free to access (http://wisplandracepillar.jic.ac.uk/) (accessed on 16 November 2022). Thus, a key aim of this study was to determine the extent of natural variation in starch digestibility of the c.Watkins lines and in modern elite UK varieties (bread, biscuit, and animal feed varieties recommended on the UK Agriculture and Horticulture Development Board Recommended List, AHDB).

Despite the availability of diverse wheat germplasm resources like the c.Watkins collection, forward screening approaches for starch digestibility have been limited due to the lack of informative, accurate, and efficient phenotyping methods. Screening based on amylose content, which has a positive association with resistant starch content, can identify lines with high levels of resistant starch but cannot identify factors beyond amylose content that may cause resistance to digestion [27,28]. Only a few studies have developed methods to directly screen large populations for starch digestibility and these have focused on analyzing purified starch [29]. However, other components of the wheat flour matrix and processing (e.g., starch retrogradation) could potentially impact starch digestibility [30–32]. To our knowledge, no previous studies have screened flour samples of wheat germplasm collections.

To facilitate rapid assessment of starch digestibility in the c.Watkins wheat landraces, an in vitro single-enzyme system was utilized (Figure 1, study workflow) [30,33,34]. This system has proved useful for measuring starch digestibility in mechanistic studies and early-stage food product development and produces results which are well correlated with human glycaemic responses to foods [33,35]. Moreover, compared to other in vitro digestion methods, the single-enzyme system has advantages such as fewer steps, low consumables, and non-specialized equipment which support its adaptation to a high-throughput assay (HTA). Here, we describe the development of a HTA for measuring starch digestibility based on the single-enzyme system [33], which utilizes a 96-sample format for simultaneous starch digestion analysis over a 90 min period. The HTA allows smaller amounts of flour to be analyzed in a standardised 96 well-format using only a thermomixer compared to a cabinet incubator, tube rotator and water bath. These modifications were made to tailor the method for screening large wheat germplasm collections accurately and efficiently. Using standard samples of wheat and maize starch, we validated the assay by comparing starch digestibility profiles to those produced by the single-enzyme system protocol reported in Edwards et al., 2019 [33]. The HTA was then used to screen for the first time processed flour samples of the entire c.Watkins collection [22] as well as elite UK varieties (representing commercial wheat lines for bread, biscuit, and animal feed), which revealed natural phenotypic variation for starch digestibility.

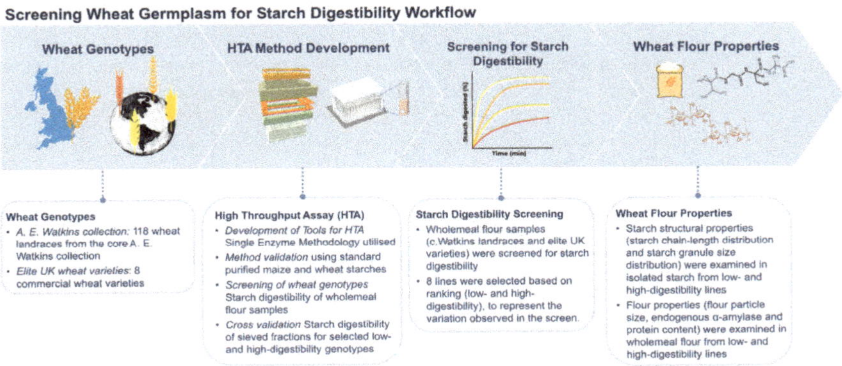

Figure 1. Study workflow for screening starch digestibility of wheat germplasm.

2. Materials and Methods

2.1. Chemicals

Chemicals used in this study: PercollTM (17-0891-01, GE Healthcare), 4-Hydroxybenzh ydrazide (PAHBAH) (5351-23-5), TRIS (77-86-1), EDTA (60-00-4), SDS (151-21-3), DTT (3483-12-3), Phosphate buffered saline (PBS) (P4417-100), Sodium Carbonate (497-19-8), DMSO (67-68-5), maltose (6363-53-7), sodium hydroxide (1310-73-2) and α-amylase (DFP Treated, Type I-A, saline suspension, 647-015-00-4) were purchased from Sigma-Aldrich Company Ltd., Poole, UK.

2.2. High-Throughput Starch Digestibility Assay

A graphical scheme of the HTA assay is presented in Figure 2. Starch digestion assays were carried out on samples that were gelatinized and cooled to accelerate retrogradation following a protocol by Edwards [33], modified for screening a large number of samples. Wholemeal flour samples were weighed (6 mg) and transferred into a deep well plate (96/1000 µL, Eppendorf, Stevenage, UK). Phosphate buffered saline (600 µL, PBS, pH 7.4) was added to each sample with a 1 mm glass ball to improve mixing. The deep well plate was sealed and secured with a Cap-mat (96-well, 7 mm, Round Plug, Silicone/PTFE) and added to a preheated thermal mixer (80 °C) with a 96 SmartBlock™ DWP 1000 n attachment (Thermomixer C, Eppendorf Ltd., Stevenage, UK) for 15 min at 1500 rpm to gelatinize the starch, then cooled at 4 °C for 21 h to accelerate retrogradation. The plate was then briefly spun (100 g for 1 min) to collect condensed liquid on the Cap-mat and placed on the thermal mixer for 30 min at 37 °C, 1600 rpm. Time zero samples (50 µL) were collected using a 12-multichannel pipette and transferred into a 2 mL deep well plate containing 1.95 mL of stop solution (17 mM $NaCO_3$). Digestion was started by adding pancreatic α-amylase suspended in PBS targeting 2 U/mL activity into the samples. Enzyme activity was determined by applying the starch digestibility assay on gelatinized potato starch and obtaining the linear rate of maltose release every 3 min (mg/mL). Aliquots (50 µL) were then taken after 6, 12, 18, 24, 40, and 90 min from the onset of digestion and transferred to the stop solution.

The stopped reactions were then centrifuged at 4000× *g* for 5 min to avoid transferring any starch remnants, and 50 µL of the supernatant was transferred to a new deep well plate; 50 µL of maltose standards (5–1000 µM) were also added to the plate. The PAHBAH reducing end assay was used to quantify the reducing ends released [36]. Briefly, 0.5 mL of freshly prepared reagent (2 g of p-hydroxybenzoic acid hydrazide dissolved in 38 mL of 0.5 M HCl and 360 mL of 0.5 M NaOH) was added to each sample. The deep-well plate was held at 100 °C for 7 min and then placed in an ice bath for 10 min. Samples were then transferred to a microplate, and absorbance was measured at 405 nm using a microplate reader (Bio-Rad Benchmark Plus, Waukegan, IL, USA). Reducing sugars were expressed as maltose equivalents, using a standard curve of maltose standards (5–1000 μM) from each sample plate. Starch digestibility (%) was expressed according to the single-enzyme system; each timepoint's maltose equivalents were corrected by subtracting the baseline maltose (time zero) and then divided by the maltose equivalent of total starch. Four technical replicates were used, each carried out on a different day.

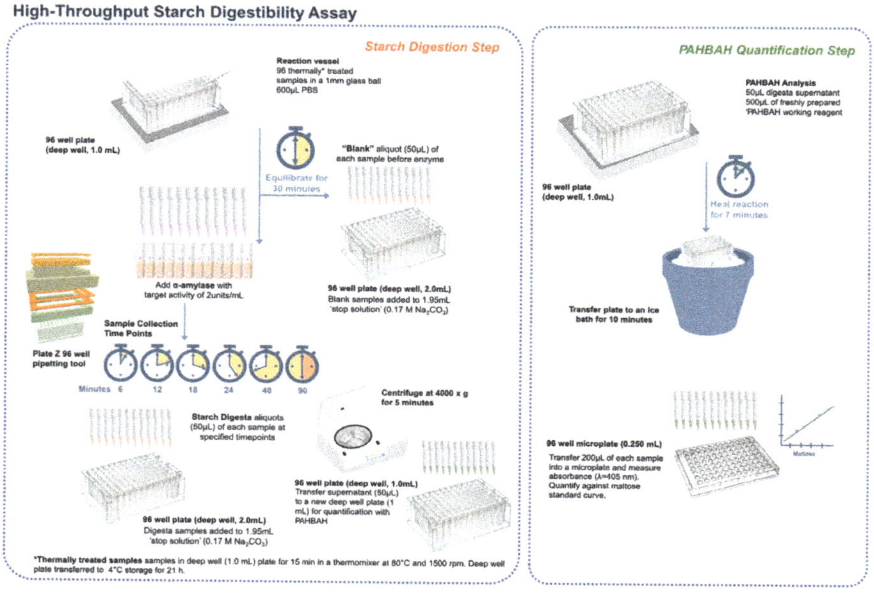

Figure 2. Principle of the high-throughput in vitro starch digestibility assay using a single enzyme system. For amylolysis, a known enzyme-substrate ratio is used, and starch is hydrolysed by porcine pancreatic α-amylase to produce reducing sugars. During amylolysis, aliquots are transferred to a 'stop solution' at predetermined time points to inactivate amylase activity. The reducing sugar concentration is quantified using a colorimetric p-hydroxybenzoic acid hydrazide (PAHBAH) assay and maltose standards [36]. The portion of starch digested for each timepoint is calculated based on reducing sugars and is then displayed against time.

2.3. *Validation of the High-Throughput Starch Digestibility Assay*

Starch digestion profiles from the HTA and the single-enzyme system were compared using standards of purified starch from standard maize, waxy maize, and high-amylose maize (purchased from Merck, formerly Sigma-Aldrich, Darmstadt, Germany), standard wheat starch (purchased from Merck, formerly Sigma-Aldrich, Darmstadt, Germany), and two high-amylose starches: sbeII and ssIIIa, previously characterized, respectively, by Corrado et al. and Fahy et al. [2,21]. Starch samples were aliquoted to make 5.4 mg of starch/mL of PBS, and α-amylase activity was adjusted at 2 U/mL. The thermal treatment procedure was followed as described above for the starch digestibility HTA. Three runs were obtained for both protocols over different days. For each run, six replicates per starch sample were placed randomly in the 96-well plate for the HTA, and two replicates were used for the standard protocol. For each run, starch samples and enzyme solution were prepared in stock and aliquoted for use in both the HTA and single-enzyme system.

2.4. *Field Trial Design*

Grains for the c.Watkins lines were ordered from the Germplasm Resources Unit (John Innes Centre in Norwich, UK) using the publicly accessible SeedStor system https://www.seedstor.ac.uk/, (accessed on 16 November 2022); permission to use the materials for research purposes was obtained. The 118 c.Watkins lines were grown in Autumn 2018 in 1 m^2 plots (one plot per line, except for the low yield lines) at Church Farm, Norfolk UK (52°37′49.2″ N 1°10′40.2″ E) using standard agronomic practices (Supplemental Figure S1). Based on yield data from previous years, lines with a lower yield performance were grown in duplicate or triplicate (to ensure production of sufficient grains), and grains were pooled for analysis (Supplemental Figure S2). Seeds from elite varieties of bread, biscuit, and animal feed commercial groups of the UK AHDB Recommended List (Cougar, Crusoe, Dickens, Diego, Myriad, Paragon, Santiago, Skyfall) were kindly provided by Brendan Fahy [37] (ahdb.org.uk) (accessed on 16 November 2022). Elite varieties were grown in 2013 in plots (one per genotype) of 1.5 m^2 at Morley Farm, Norfolk, UK (52°33′15.57″ N 1°10′58.956″ E).

2.5. Milling and Sieving

Grains from the c.Watkins and elite varieties were coarsely milled in a cyclone mill fitted with a 0.5 mm screen (UDY Corporation). Milled samples were passed through a 0.3 mm sieve (Endecotts Limited, London, UK) to produce 'wholemeal' flour samples, and selected lines were passed through a 0.053 mm sieve to produce 'sieved' flour samples. The flour samples were kept in a vacuum desiccator for five days before analysis.

2.6. Starch Isolation

Starch was isolated using an adapted method reported in Hawkins, et al. [38]. Wheat flour samples were resuspended with water and filtered through a 100 µm cell strainer (BD Falcon #352360). Samples were then centrifuged at 3000× *g* for 5 min, and the pellets were resuspended in 2 mL of water. The starch suspensions were then overlayed into a Percoll solution (90% *v*/*v*) and centrifuged at 2500× *g* for 15 min to remove cell walls and proteins. The recovered starch pellets were washed with 1 mL buffer (50 mM Tris-HCl, pH 6.8; 10 mM EDTA; 4% SDS; and 10 mM DTT), transferred into a 2 mL tube and left to incubate for 5 min. The starch suspension was then centrifuged at 4000× *g* for 1 min. The pellets were recovered, and the washing procedure was repeated once more. The pellets were then washed three times with 1 mL of water, then once with 100% ethanol. Samples were then kept one day in the fume hood, followed by five days in a desiccator containing silicon dioxide prior to analysis.

2.7. Total Starch Assay

Wholemeal flour samples were weighed (~8 mg) and transferred into a deep well plate (96/1000 µL, Eppendorf), each well containing 20 µL of DMSO and a 3 mm glass ball to improve mixing. The plate was mixed for 5 min at 1600 rpm to disperse the samples before adding 500 µL of a thermostable α-amylase to each sample. The thermostable α-amylase was solubilized at 1:30 (*v*/*v*) in 100 mM sodium acetate buffer, pH 5.0 (Total Starch hexokinase kit, AOAC Method 996.1 1; Megazyme, Bray, IE). The deep well plate was sealed and secured with a Cap-mat (96-well, 7 mm, Round Plug, Silicone/PTFE). Samples were heated at 90 °C for 10 min at 1600 rpm using a thermal mixer with a 96 SmartBlock™ DWP 1000 n attachment (Thermomixer C, Eppendorf Ltd., Stevenage, UK). Total starch content was determined using a Total Starch HK kit (Total Starch hexokinase kit, AOAC Method 996.1 1; Megazyme, Bray, IE, Wicklow, Ireland) following manufacturer instructions, except volumes of reagents were scaled down by a factor of 10.

2.8. In Depth Analysis of Selected High- and Low-Digestibility Lines

Methods for particle size analysis of flour and starch, size exclusion chromatography, protein content and endogenous α-amylase are available in Supplementary Materials.

2.9. Tools

During optimization of the HTA, a low-cost 3D-printed pipetting tool was developed, which allowed for manageable weighing and transferring of samples into 96-sample deep well plates and improved speed and control of pipette aspiration (Plate Z). The PLZ 3D design is available to download and print for free (https://www.hackster.io/386082/high-throughput-pipetting-plate-z-bde2c7) (accessed on 16 November 2022).

2.10. Data Analysis

Statistical analyses and graphs were produced using RStudio (R version 4.2.1, Posit Software, Boston, MA, USA). Datasets of the validation of the in vitro starch digestibility HTA were analysed using the packages lme4 (v 1.1-30) and lmerTest (v 3.1-3) for a mixed model fit [39–41]. Plots were made using the ggplot2 package (v 3.3.6) [42].

For validation of the HTA, the methods were compared by plotting the estimated starch digestion profiles from the HTA and the single-enzyme system (protocol from Edwards et al., 2019 [33]) and by comparing the estimated starch digested at 90 min. The bias (difference in estimates between methods for the same material) and variation between runs (technical variation) in starch digested at 90 min was estimated using linear mixed models, including the type of starch as a fixed effect and sample batch as a random effect. The variances were estimated using separate models for each method, so that the variability of each method could be compared, while the bias was estimated using a single joint model including all data points, with 'method' corresponding to an additional fixed effect.

A linear mixed model including line as a fixed effect and experimental run as a random effect was used for analysis of in vitro starch digestibility of the c.Watkins lines and elite varieties. Marginal means with standard errors (calculated using a pooled standard deviation) are plotted for each line.

The correlation between starch digestibility and total starch was estimated using linear regression. All values reported represent the mean, and the number of replicates and variance metrics are specified in the description of the corresponding figure and Supplementary Datasets.

For complete details of analyses, all data and analysis code is available as Supplementary Materials.

3. Results and Discussion

3.1. Screening of c.Watkins Landraces and Elite Varieties

3.1.1. In Vitro Starch Digestibility

The aim of this experiment was to determine the extent of natural variation in starch digestibility of the c.Watkins collection and compare it with elite UK varieties representing each commercial group (qualities of bread, biscuit and animal feed).

Results of the HTA revealed a wide range of variation in starch digestibility among c.Watkins landraces; starch digestibility profiles formed a gradient of low- to high-digestibility rather than two distinct groups, as expected from complex traits (Figure 3A shows the individual trajectories of starch digested over time, Figure 3B compares the starch digested for each line at 90 min). For the c.Watkins lines, the levels of starch digested at 90 min ranged between 9.7–31.6% (at 6 min), 13.2–35% (at 12 min), 14.8–37.2% (at 18 min), 16.2–36.1% (at 24 min), 19–37.8% (at 40 min), and 23.5–39.9% (Figure 3B). The residual variance from the mixed effect model, measuring the variability between technical replicates from the same sample in the same run, was 3.7 percentage points. There was no statistical evidence for a relationship between the plot location in the field and the starch digestibility at 90 min, as no trend was observed between field rows and columns (Supplemental Figure S3).

Elite UK varieties showed less variation and greater starch digestibility compared to most c.Watkins lines, although a direct comparison is confounded by the different growing conditions between the two groups (Figure 3A). For example, the levels of starch digested for the elite varieties ranged between 26.8–36.4% (at 6 min), 29.4–40.4% (at 12 min), 30.6–43.8% (at 18 min), 30–43.4% (at 24 min), 31.3–43.4% (at 40 min) and 31.2–43.5% (at 90 min) (Figure 3B).

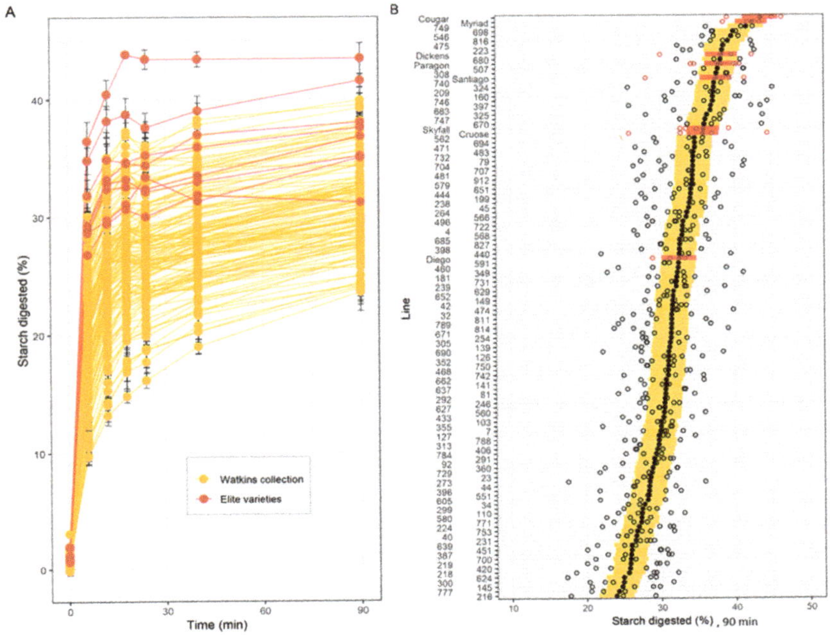

Figure 3. Starch digestibility and total starch of c.Watkins landraces (yellow) and elite varieties (red) of wholemeal flour. (**A**). Starch digested (%) for c.Watkins (yellow) and elite varieties (red). Points and lines represent the mean and standard error from between 3 and 6 technical replicates per line. (**B**). Starch digested (%) at 90 min. Individual data points from technical replicates are shown in white dots, and marginal mean values from a mixed effects model with line as a fixed effect and experimental run as a random effect are shown in black (c.Watkins) and red dots (elite). The standard error estimated from the model is displayed as yellow (c.Watkins) and red (elite) bars. Values represent mean ± SE of $n \geq 3$ replicates.

Our results demonstrate that natural phenotypic variation for starch digestibility exists in the c.Watkins collection (representing the genetic diversity of the entire Watkins collection, 826 lines) as well as in elite varieties representing commercial groups of bread, biscuit and animal feed groups, and so there is potential for identifying underlying genetic diversity. The variation observed among the c.Watkins lines was greater than among the elite varieties screened, which is consistent with the higher levels of genetic diversity previously reported for the c.Watkins collection compared to modern European bread wheat varieties, as well as increased phenotypic variation for agronomically desirable traits, such as stripe, leaf, and stem rust resistance, as well as grain surface area and grain width [22–26]. It is important to note that a limitation of our study was that the grain samples of c.Watkins lines and elite varieties were not produced in the same field trial, and each a single plot per line, so it is possible that environmental conditions between and within fields could influence the average starch digestibility levels and the variation observed; this will be important to consider in future trials. With the availability of structured germplasm panels (nested association mapping populations) and genotypic data [43], there is now potential for use of the HTA to investigate the genetic factors underlying the variation in starch digestibility observed through approaches like QTL analysis which will aid in identification of underlying candidate genes and development of tools for marker assisted-selection. Furthermore, once lines are identified, digestion of starch and other nutrients in wheat foods can be explored using more extensive in vitro models of digestion in the upper gastrointestinal track such as the standardized INFOGEST protocol, based on an international consensus by COST INFOGEST network [44].

3.1.2. Total Starch

Total starch (TS) content of wholemeal flour varied significantly for c.Watkins landraces (43 ± 3.3 g/100 flour to 61 ± 2.4 g/100 flour, mean ± SE, $p \leq 0.001$), and the elite varieties (46 ± 3.6 g/100 flour to 61 ± 0.9 g/100 flour, mean ± SE, $p \leq 0.05$). The elite variety Diego had the highest TS content (61 ± 0.9 g/100 flour, mean ± SE), and the c.Watkins line 651 had the lowest (43 ± 3.3 g/100 flour, mean ± SE). Most of the samples had a TS content between 47 to 57 g/100 flour. Linear regression analysis showed that total starch content only weakly correlated ($R^2 = 0.0108$) with starch digested at 90 min which suggests that the differences in total starch content do not explain the variation observed for starch digestibility (Supplemental Figure S4).

3.1.3. Analysis of Low- and High-Digestibility Lines

To gain further insight into factors influencing the difference in starch digestibility a subset of c.Watkins lines and elite varieties were selected, based on their starch digestibility profiles (high- vs. low-digestibility) to measure starch digestibility in sieved flour (to obtain smaller particle size fractions) and purified starch.

Three low-digestibility lines: 777 (WATDE0111), 216 (WATDE0025), and 639 (WATDE0083); and five high-digestibility lines: including two c.Watkins landraces 816 (WATDE0117), 308 (WATDE0042) and the three UK elite varieties Myriad, Dickens, and Paragon were selected for further analysis. The selection was based on the starch digestibility HTA results, specifically the ranking (low to high), to represent the variation observed in the screen.

Results presented in Figure 4 show that starch digestibility profiles differed considerably for wholemeal flour, sieved flour, and purified starch, suggesting that other flour components besides starch likely may contribute to the variation in starch digestibility observed. For wholemeal flour (<0.3 mm), starch digestibility profiles of selected 'low-'and 'high-digestibility' lines grouped separately where the high-digestibility samples varied between 37.6–41.6% and low-digestibility samples varied between 23.5–24.3% (at 90 min) (Figure 4A). For sieved flour (<0.05 mm), the digestibility of two low digestibility-lines (216 and 639) increased significantly ($p < 0.05$) compared to the corresponding wholemeal samples, and while the third low-line (777) remained low (Figure 4B). Purified starch samples showed a greater extent of starch digested at 90 min compared to wholemeal and sieved flour fractions (Figure 4C). Moreover, there were no differences between the high- and low-digestibility groups in purified starch samples. For example, high-digestibility lines differed between 46.9–52.3% and low-digestibility lines varied between 47.6–49.2% (at 90 min).

Finally, these selected lines were also analyzed for starch structural properties (starch chain-length distribution and starch granule size distribution) and flour properties (flour particle size, endogenous α-amylase and protein content). These results are described below. In summary, while differences between the lines themselves were evident for many properties, none of these factors significantly correlated with starch digestibility in this sample (Supplemental Table S1 and Supplemental Figures S5–S7).

Starch molecular structure and composition have been shown to influence its digestibility; thus, starch fine molecular structure and starch granule size-distribution was examined in the selected low- and high-digestibility lines. Starch chain-length distributions showed significant differences between lines in the proportion of long and short chains of amylopectin (AP) and amylose (AM) ($p < 0.001$) (Supplemental Figure S5 and Table S1). However, the overall AM:AP ratio was not significant. Starch granule size distribution of purified starch using a Coulter counter revealed significant variation among starch granule diameter ($p < 0.001$), whereas minor differences were observed for the overall volume of A and B granules (Supplemental Table S1). Low-digestibility lines showed the greatest variation in B-granule diameter, with 6 ± 0.1 µm SE for line 777 to 8.4 ± 0.2 µm SE for line

216 (40% greater). The diameter of B-granules in high-digestibility lines varied less; 816 had the highest (6.9 ± 0.4 μm SE), and 308 had the lowest (6.2 ± 0.1 μm SE). In general, similar trends were also observed for A-granule diameter, except that low-digestibility line 216 had significantly larger A-granules. The diameter of A-granules from low digestibility lines ranged from 18.2 ± 0.2 μm SE (lines 777 and 639) to 20.1 ± 0.2 μm SE (line 216). There was also significant variation in A-granule diameter among high-digestibility lines; 308 had the smallest (17.7 ± 0.04 μm SE) and Paragon the largest (20.1, ± 0.2 μm SE). Overall, we observed differences in chain-length distribution profiles and starch granule size distribution of the selected lines but the there was no correlation of either factor to starch digestibility of the wholemeal flour.

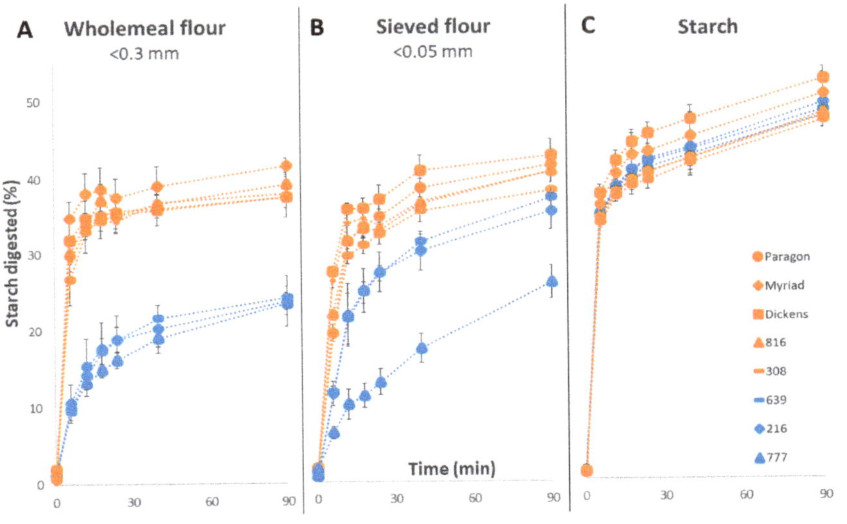

Figure 4. Starch digestibility of selected low- and high-digestibility lines measured in wholemeal flour (**A**), sieved flour (**B**) and purified starch (**C**) using the HTA. Samples in blue represent the low-digestibility lines, and samples in orange represent the high-digestibility lines. Values represent mean ± SE of $n = 3$ replicates.

Starch digestibility has also been shown to be affected by the particle size of wheat flour [17]. Particle size analysis of wholemeal flour from selected low- and high-digestibility lines showed no major variation (Supplemental Figure S6). There was a statistically significant difference ($p < 0.003$) in the particle size of wholemeal flour; line 639 had more particles in the range of 14–20 μm whereas Paragon showed a slightly smaller number of particles in that range. Micrographs of wholemeal flour (produced using scanning electron microscopy) showed aggregate formations ≥ 120μm in all lines apart from the low-digestibility line 639, which only had particles smaller than 120μm (Supplemental Figure S7). This data suggests that the milling efficiency was similar in all lines apart from 639.

Elevated amounts of endogenous α-amylase in cereals could trigger early starch amylolysis in the endosperm and thus significantly affect starch digestibility [45,46]. The activity of endogenous α-amylase in low- and high-digestibility lines differed significantly ($p < 0.05$), (Supplemental Table S1), however values were within a normal range (<0.2 Ceralpha Units/g) according to prior studies (McCleary et al., 2002, Derkx and Mares, 2020). The activity of endogenous α-amylase in low-digestibility lines ranged from 0.06 ± 0.01 to 0.15 ± 0.01 Ceralpha Units/g of flour, mean ± SD, with the lowest in line 639 and the highest line 216. High-digestibility lines varied from 0.05 ± 0.01 to

0.13 ± 0.01 Ceralpha Units/g of flour, mean $\pm$ SD, Paragon being the lowest and line 308 the highest.

Grain hardness can impact starch digestibility and is mainly affected by the protein content and composition in the endosperm [47]. Thus, protein content of wholemeal flour was analysed in the selected lines. Results showed that protein content varied significantly ($p < 0.001$) (Supplemental Table S1); low-digestibility lines ranged from 14.2–18.4 g/100 flour, mean and high-digestibility lines ranged from 10.6–17.4 g/100 g flour, mean.

The different digestibility profiles observed across purification steps (wholemeal flour to sieved flour to purified starch) suggest that multiple mechanisms in the selected lines could be affecting starch digestibility and that different factors in each line could have a distinct effect. This is consistent with prior studies which have shown effects of flour characteristics and starch structure on starch digestibility in cooked wheat samples including particle size of wholemeal flour, protein content, and chain-length distribution of starch polymers [31,48,49]. An important limitation of this aspect of the study was the small number of lines selected for this analysis severely limiting the power to detect correlations of phenotypes between lines; prior work by Wang et al. [29] identified starch properties influencing starch digestibility in a screen of 224 wheat starches. Nevertheless, our study suggests that a combination and interaction of many factors are required to achieve low starch digestibility profiles in flour, so using a starch digestibility HTA for screening or selection approaches may be more efficient and informative than selection based solely on underlying factors such as starch molecular structure.

3.2. Establishment and Validation of a High-Throughput Starch Digestibility Assay

The aim of this experiment was to validate and assess the reliability of the 96-sample format starch digestibility assay using established starch standards of maize and wheat by comparing the HTA with the single-enzyme system protocol presented in Edwards et al., 2019 [33].

The HTA (detailed in the Section 2) was scaled from a 6-tube format to a 96-well plate format which required optimizations for maintaining even heat distribution during starch digestions and PAHBAH assays, for the mixing ability of samples, and for recovering adequate sample volumes for analysis. We also developed new tools to allow faster sample handling for sample preparation and pipetting (Plate Z, available free to download from hackster.io, (accessed on 16 November 2022). A graphical scheme is presented in the Supplemental Materials highlighting the changes made to develop the HTA (Supplementary Figure S8).

Digestion profiles of maize and wheat starch standards were comparable to profiles generated using the single-enzyme system protocol in Edwards et al., 2019 [33] (Figure 5A). For example, the difference in the average estimates at 90 min (Figure 5B) from the two methods was −0.72 percentage points (not statistically significant), and the technical variation observed at 90 min within the runs was 2.1 percentage points in the HTA and 2.2 percentage points in the single-enzyme system protocol, suggesting that the HTA is not biased and is no less reliable.

There were no significant differences in the percent of starch digested at all the time points measured between the single-enzyme system protocol and the HTA. Thus, the HTA provided comparisons between samples that were accurate and reproducible, and the reliability of the assay was sufficient for use as a screening tool to aid in the selection of low- and high-digestibility samples.

The variance using wholemeal flour (in the c.Watkins screen, Section 3.1.1) was higher than the variance using starch standards in the validation experiment which could be due to the higher variability in wholemeal flour composition compared to purified starch samples. The number of technical replicates needed to estimate starch digestibility to a given precision can be calculated from residual variance.

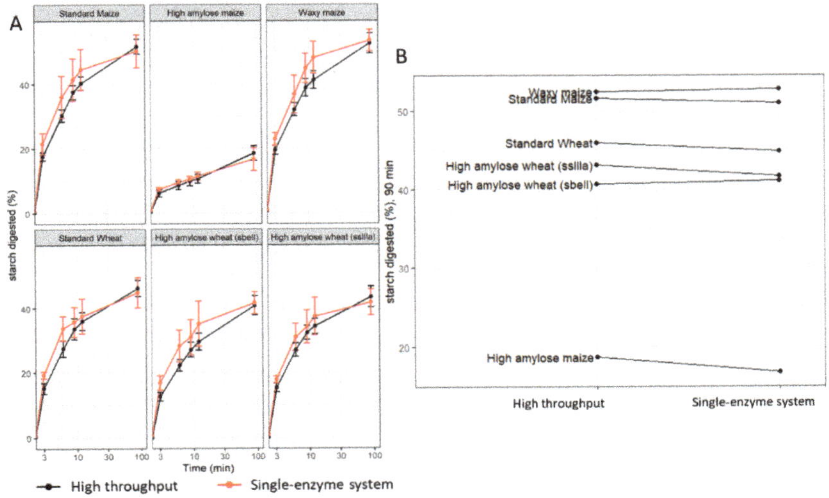

Figure 5. Comparison of the HTA to the single-enzyme system protocol reported in Edwards et al., 2019. (**A**) Digestibility profiles of purified maize and wheat starch produced by the HTA (black) and the single-enzyme system (red). (**B**) Starch digested (%) at 90 min, ranking comparison of the HTA (left) to the single-enzyme system (right). Values represent the mean ± SD of n = 18 replicates for the HTA and n = 6 for the single-enzyme system.

Recent studies have reported improved methods to increase the throughput of starch digestibility assays. Wang et al. [29] developed a 96-sample format assay to screen starch from a wheat MAGIC population and used a multifactorial analysis to identify the most influential factors for starch digestibility (starch granule size distribution, amylopectin chain length distribution, amylose content and endogenous α-amylase activity). Other studies have also aimed to increase the throughput of starch characterization techniques however, the starch properties analyzed may or may not have a direct impact on starch digestibility [50,51]. It is important to consider that other components of wheat flour and factors like processing are likely to have a major impact on starch digestibility thus, determining starch digestibility on processed flour samples may better represent what people may consume in wheat-based foods [29,30,33,53]. Some progress has been made for other crops like rice; Toutounji et al. [52] developed a 15-sample format assay for screening starch digestibility of cooked rice grains with potential to allow analysis of 60 samples per day (if 15 samples are prepared every two hours). In this method, samples were handled individually with a single pipette, which was a limiting factor for the number of samples that could be processed simultaneously. Furthermore, the assay was performed using a set of 8 samples and this assay still needs to be validated on a larger sample set. The HTA assay described here addresses some of these key limitations presented by prior assays namely the type of sample analyzed (hydrothermally processed flour) and sample size (96-sample format). For future work, it will be important to consider improving the efficiency of upstream steps, including milling, sieving, and weighing samples which present additional bottlenecks.

4. Conclusions

In this study we identified wide variation in starch digestibility among the c.Watkins lines which could be used to recover valuable alleles that were lost during modern breeding with the aim of improving wheat nutritional quality. Starch digestibility profiles of sieved flour and purified starch from selected lines suggested that the differences observed in starch digestibility are likely due to multiple factors in the flour matrix and are not limited to starch structural properties. Previously, strategies to reduce starch digestibility in wheat have mainly focused on reverse genetic approaches, but with new tools like the HTA there is potential to discover other useful sources of variation to support development of wheat varieties with improved health benefits. Furthermore, future work to determine starch digestibility in white flour samples will be useful for identifying sources of reduced starch digestibility for applications in foods made with refined flour.

Supplementary Materials: The following supporting information can be downloaded at: https://www.mdpi.com/article/10.3390/foods12020266/s1, Supplemental.docx, data sets of starch digestibility and total starch, and all statistical analysis code. Additional references cited in Supplementary Materials: [53,54].

Author Contributions: P.Z.: Conceptualization, investigation, methodology, project administration, data curation, and formal analysis, writing—original draft preparation, writing—review and editing, G.M.S.: methodology, data curation, visualization, formal analysis, writing—review and editing, J.H.A.-J.: investigation, methodology, writing—review and editing, F.J.W.: conceptualization, methodology, supervision, writing—review and editing, M.P.: investigation, methodology, writing—review and editing, S.G.: conceptualization, methodology, supervision, writing—review and editing, D.S.: conceptualization, investigation, methodology, supervision, writing—review and editing, B.A.H.: conceptualization, funding acquisition, methodology, supervision, writing—review and editing. All authors have read and agreed to the published version of the manuscript.

Funding: This research was funded by the UKRI Biotechnology and Biological Sciences Research Council Norwich Research Park Biosciences Doctoral Training Partnership [grant number BB/M011216/1] (PZ) and the UKRI Biotechnology and Biological Sciences Research Council (BBSRC) Institute Strategic Programme 'Food Innovation and Health' (grant number BB/R012512/1) and its constituent projects BBS/E/F/000PR10343 (Theme 1, Food Innovation) and BBS/E/F/000PR10345 (Theme 2, Digestion in the Upper GI Tract) and the BBSRC Core Capability Grant BB/CCG1860/1 awarded to Quadram Institute Bioscience; the BBSRC Institute Strategic Programme Grants 'Molecules from Nature'—Crop Quality BBS/E/J/000PR9799 awarded to the John Innes Centre and 'Designing Future Wheat' BB/P016855/1 awarded to the John Innes Centre.

Data Availability Statement: Data sets of starch digestibility and total starch and all statistical analysis code are included as Supplementary Materials.

Conflicts of Interest: The authors declare no conflict of interest. The funders had no role in the design of the study; in the collection, analyses, or interpretation of data; in the writing of the manuscript; or in the decision to publish the results.

References

1. Blaak, E.E.; Antoine, J.M.; Benton, D.; Björck, I.; Bozzetto, L.; Brouns, F.; Diamant, M.; Dye, L.; Hulshof, T.; Holst, J.J.; et al. Impact of postprandial glycaemia on health and prevention of disease. *Obes. Rev.* **2012**, *13*, 923–984. [CrossRef] [PubMed]
2. Corrado, M.; Ahn-Jarvis, J.H.; Fahy, B.; Savva, G.M.; Edwards, C.H.; Hazard, B.A. Effect of high-amylose starch branching enzyme II wheat mutants on starch digestibility in bread, product quality, postprandial satiety and glycaemic response. *Food Funct.* **2022**, *13*, 1617–1627. [CrossRef] [PubMed]
3. Belobrajdic, D.P.; Regina, A.; Klingner, B.; Zajac, I.; Chapron, S.; Berbezy, P.; Bird, A.R. High-amylose wheat lowers the postprandial glycemic response to bread in healthy adults: A randomized controlled crossover trial. *J. Nutr.* **2019**, *149*, 1335–1345. [CrossRef] [PubMed]
4. Hughes, R.L.; Horn, W.H.; Finnegan, P.; Newman, J.W.; Marco, M.L.; Keim, N.L.; Kable, M.E. Resistant Starch Type 2 from Wheat Reduces Postprandial Glycemic Response with Concurrent Alterations in Gut Microbiota Composition. *Nutrients* **2021**, *13*, 645. [CrossRef] [PubMed]
5. Shewry, P.R.; Hey, S.J. The contribution of wheat to human diet and health. *Food Energy Secur.* **2015**, *4*, 178–202. [CrossRef] [PubMed]
6. Hazard, B.; Trafford, K.; Lovegrove, A.; Griffiths, S.; Uauy, C.; Shewry, P. Strategies to improve wheat for human health. *Nat. Food* **2020**, *1*, 475–480. [CrossRef]
7. Shewry, P.R.; Hawkesford, M.J.; Piironen, V.; Lampi, A.-M.; Gebruers, K.; Boros, D.; Andersson, A.A.; Åman, P.; Rakszegi, M.; Bedo, Z. Natural variation in grain composition of wheat and related cereals. *J. Agric. Food Chem.* **2013**, *61*, 8295–8303. [CrossRef]
8. Delcour, J.A.; Hoseney, R.C. *Principles of Cereal Science and Technology*, 3rd ed.; AACC International: St. Paul, MN, USA, 2010.
9. Wang, S.; Copeland, L. Molecular disassembly of starch granules during gelatinization and its effect on starch digestibility: A review. *Food Funct.* **2013**, *4*, 1564–1580. [CrossRef]

10. Chung, H.-J.; Lim, H.S.; Lim, S.-T. Effect of partial gelatinization and retrogradation on the enzymatic digestion of waxy rice starch. *J. Cereal Sci.* **2006**, *43*, 353–359. [CrossRef]
11. Parada, J.; Aguilera, J.M. In vitro digestibility and glycemic response of potato starch is related to granule size and degree of gelatinization. *J. Food Sci.* **2009**, *74*, E34–E38. [CrossRef]
12. Wang, S.; Li, C.; Copeland, L.; Niu, Q.; Wang, S. Starch retrogradation: A comprehensive review. *Compr. Rev. Food Sci. Food Saf.* **2015**, *14*, 568–585. [CrossRef]
13. Corrado, M.; Cherta-Murillo, A.; Chambers, E.S.; Wood, A.J.; Plummer, A.; Lovegrove, A.; Edwards, C.H.; Frost, G.S.; Hazard, B.A. Effect of semolina pudding prepared from starch branching enzyme IIa and b mutant wheat on glycaemic response in vitro and in vivo: A randomised controlled pilot study. *Food Funct.* **2020**, *11*, 617–627. [CrossRef] [PubMed]
14. Hazard, B.; Zhang, X.; Naemeh, M.; Hamilton, M.K.; Rust, B.; Raybould, H.E.; Newman, J.W.; Martin, R.; Dubcovsky, J. Mutations in durum wheat SBEII genes affect grain yield components, quality, and fermentation responses in rats. *Crop Sci.* **2015**, *55*, 2813–2825. [CrossRef] [PubMed]
15. Schönhofen, A.; Hazard, B.; Zhang, X.; Dubcovsky, J. Registration of common wheat germplasm with mutations in SBEII genes conferring increased grain amylose and resistant starch content. *J. Plant Regist.* **2016**, *10*, 200–205. [CrossRef] [PubMed]
16. Anugerah, M.P.; Faridah, D.N.; Afandi, F.A.; Hunaefi, D.; Jayanegara, A. Annealing processing technique divergently affects starch crystallinity characteristic related to resistant starch content: A literature review and meta-analysis. *Int. J. Food Sci. Technol.* **2022**, *57*, 2535–2544. [CrossRef]
17. Edwards, C.H.; Grundy, M.M.; Grassby, T.; Vasilopoulou, D.; Frost, G.S.; Butterworth, P.J.; Berry, S.E.; Sanderson, J.; Ellis, P.R. Manipulation of starch bioaccessibility in wheat endosperm to regulate starch digestion, postprandial glycemia, insulinemia, and gut hormone responses: A randomized controlled trial in healthy ileostomy participants. *Am. J. Clin. Nutr.* **2015**, *102*, 791–800. [CrossRef]
18. López-Barón, N.; Gu, Y.; Vasanthan, T.; Hoover, R. Plant proteins mitigate in vitro wheat starch digestibility. *Food Hydrocoll.* **2017**, *69*, 19–27. [CrossRef]
19. Botticella, E.; Sestili, F.; Sparla, F.; Moscatello, S.; Marri, L.; Cuesta-Seijo, J.A.; Falini, G.; Battistelli, A.; Trost, P.; Lafiandra, D. Combining mutations at genes encoding key enzymes involved in starch synthesis affects the amylose content, carbohydrate allocation and hardness in the wheat grain. *Plant Biotechnol. J.* **2018**, *16*, 1723–1734. [CrossRef]
20. Schonhofen, A.; Zhang, X.Q.; Dubcovsky, J. Combined mutations in five wheat STARCH BRANCHING ENZYME II genes improve resistant starch but affect grain yield and bread-making quality. *J. Cereal Sci.* **2017**, *75*, 165–174. [CrossRef]
21. Fahy, B.; Gonzalez, O.; Savva, G.M.; Ahn-Jarvis, J.H.; Warren, F.J.; Dunn, J.; Lovegrove, A.; Hazard, B.A. Loss of starch synthase IIIa changes starch molecular structure and granule morphology in grains of hexaploid bread wheat. *Sci. Rep.* **2022**, *12*, 1–14. [CrossRef]
22. Wingen, L.U.; Orford, S.; Goram, R.; Leverington-Waite, M.; Bilham, L.; Patsiou, T.S.; Ambrose, M.; Dicks, J.; Griffiths, S. Establishing the A. E. Watkins landrace cultivar collection as a resource for systematic gene discovery in bread wheat. *Theor. Appl. Genet.* **2014**, *127*, 1831–1842. [CrossRef] [PubMed]
23. Bansal, U.K.; Forrest, K.L.; Hayden, M.J.; Miah, H.; Singh, D.; Bariana, H.S. Characterisation of a new stripe rust resistance gene Yr47 and its genetic association with the leaf rust resistance gene Lr52. *Theor. Appl. Genet.* **2011**, *122*, 1461–1466. [CrossRef]
24. Bansal, U.K.; Arief, V.N.; DeLacy, I.H.; Bariana, H.S. Exploring wheat landraces for rust resistance using a single marker scan. *Euphytica* **2013**, *194*, 219–233. [CrossRef]
25. Randhawa, M.S.; Bariana, H.S.; Mago, R.; Bansal, U.K. Mapping of a new stripe rust resistance locus Yr57 on chromosome 3BS of wheat. *Mol. Breed.* **2015**, *35*, 65. [CrossRef]
26. Toor, A.K.; Bansal, U.K.; Bhardwaj, S.; Badebo, A.; Bariana, H.S. Characterization of stem rust resistance in old tetraploid wheat landraces from the Watkins collection. *Genet. Resour. Crop Evol.* **2013**, *60*, 2081–2089. [CrossRef]
27. Chen, M.-H.; Bergman, C.J.; McClung, A.M.; Everette, J.D.; Tabien, R.E. Resistant starch: Variation among high amylose rice varieties and its relationship with apparent amylose content, pasting properties and cooking methods. *Food Chem.* **2017**, *234*, 180–189. [CrossRef] [PubMed]
28. Mishra, A.; Singh, A.; Sharma, M.; Kumar, P.; Roy, J. Development of EMS-induced mutation population for amylose and resistant starch variation in bread wheat (Triticum aestivum) and identification of candidate genes responsible for amylose variation. *BMC Plant Biol.* **2016**, *16*, 1–15. [CrossRef] [PubMed]
29. Wang, Y.; Kansou, K.; Pritchard, J.; Zwart, A.B.; Saulnier, L.; Ral, J.-P. Beyond amylose content, selecting starch traits impacting in vitro α-amylase degradability in a wheat MAGIC population. *Carbohydr. Polym.* **2022**, *291*, 119652. [CrossRef]
30. Edwards, C.H.; Warren, F.J.; Milligan, P.J.; Butterworth, P.J.; Ellis, P.R. A novel method for classifying starch digestion by modelling the amylolysis of plant foods using first-order enzyme kinetic principles. *Food Funct.* **2014**, *5*, 2751–2758. [CrossRef] [PubMed]
31. Sissons, M.; Cutillo, S.; Marcotuli, I.; Gadaleta, A. Impact of durum wheat protein content on spaghetti in vitro starch digestion and technological properties. *J. Cereal Sci.* **2021**, *98*, 103156. [CrossRef]
32. Qi, K.; Yi, X.; Li, C. Effects of endogenous macronutrients and processing conditions on starch digestibility in wheat bread. *Carbohydr. Polym.* **2022**, *295*, 119874. [CrossRef] [PubMed]
33. Edwards, C.H.; Cochetel, N.; Setterfield, L.; Perez-Moral, N.; Warren, F.J. A single-enzyme system for starch digestibility screening and its relevance to understanding and predicting the glycaemic index of food products. *Food Funct.* **2019**, *10*, 4751–4760. [CrossRef] [PubMed]

34. Butterworth, P.J.; Warren, F.J.; Grassby, T.; Patel, H.; Ellis, P.R. Analysis of starch amylolysis using plots for first-order kinetics. *Carbohydr. Polym.* **2012**, *87*, 2189–2197. [CrossRef]
35. Goñi, I.; Garcia-Alonso, A.; Saura-Calixto, F. A starch hydrolysis procedure to estimate glycemic index. *Nutr. Res.* **1997**, *17*, 427–437. [CrossRef]
36. Lever, M. A new reaction for colorimetric determination of carbohydrates. *Anal. Biochem.* **1972**, *47*, 273–279. [CrossRef]
37. Fahy, B.; Siddiqui, H.; David, L.C.; Powers, S.J.; Borrill, P.; Uauy, C.; Smith, A.M. Final grain weight is not limited by the activity of key starch-synthesising enzymes during grain filling in wheat. *J. Exp. Bot.* **2018**, *69*, 5461–5475. [CrossRef]
38. Hawkins, E.; Chen, J.; Watson-Lazowski, A.; Ahn-Jarvis, J.; Barclay, J.E.; Fahy, B.; Hartley, M.; Warren, F.J.; Seung, D. STARCH SYNTHASE 4 is required for normal starch granule initiation in amyloplasts of wheat endosperm. *BioRxiv* **2021**, *230*, 2371–2386. [CrossRef]
39. Kunzetsova, A.; Brockhoff, P.; Christensen, R. lmerTest package: Tests in linear mixed effect models. *J. Stat. Softw.* **2017**, *82*, 1–26.
40. Bates, D.; Maechler, M.; Bolker, B.; Walker, S. Fitting linear mixed-effects models using lme4. *arXiv* **2014**, arXiv:1406.5823.
41. Pinheiro, J.; Bates, D.; DebRoy, S.; Sarkar, D. R Core Team. 2021. Nlme: Linear and Nonlinear Mixed Effects Models. R Package Version 3.1-153. Available online: https://cran.r-project.org/web/packages/nlme/index.html (accessed on 31 March 2022).
42. Wickham, H. Data Analysis. In *ggplot2: Elegant Graphics for Data Analysis*; Springer International Publishing: Cham, Switzerland, 2016; pp. 189–201.
43. Wingen, L.U.; West, C.; Leverington-Waite, M.; Collier, S.; Orford, S.; Goram, R.; Yang, C.Y.; King, J.; Allen, A.M.; Burridge, A.; et al. Wheat Landrace Genome Diversity. *Genetics* **2017**, *205*, 1657–1676. [CrossRef]
44. Brodkorb, A.; Egger, L.; Alminger, M.; Alvito, P.; Assunção, R.; Ballance, S.; Bohn, T.; Bourlieu-Lacanal, C.; Boutrou, R.; Carrière, F.; et al. INFOGEST static in vitro simulation of gastrointestinal food digestion. *Nat. Protoc.* **2019**, *14*, 991–1014. [CrossRef] [PubMed]
45. Newberry, M.; Zwart, A.B.; Whan, A.; Mieog, J.C.; Sun, M.; Leyne, E.; Pritchard, J.; Daneri-Castro, S.N.; Ibrahim, K.; Diepeveen, D.; et al. Does Late Maturity Alpha-Amylase Impact Wheat Baking Quality? *Front. Plant Sci.* **2018**, *9*, 1356. [CrossRef] [PubMed]
46. Derkx, A.P.; Mares, D.J. Late-maturity α-amylase expression in wheat is influenced by genotype, temperature and stage of grain development. *Planta* **2020**, *251*, 51. [CrossRef] [PubMed]
47. Morris, C.F. Puroindolines: The molecular genetic basis of wheat grain hardness. *Plant Mol. Biol.* **2002**, *48*, 633–647. [CrossRef] [PubMed]
48. Obadi, M.; Li, C.; Li, Q.; Li, X.; Qi, Y.; Xu, B. Relationship between starch fine molecular structures and cooked wheat starch digestibility. *J. Cereal Sci.* **2020**, *95*, 103047. [CrossRef]
49. Lin, S.; Gao, J.; Jin, X.; Wang, Y.; Dong, Z.; Ying, J.; Zhou, W. Whole-wheat flour particle size influences dough properties, bread structure and in vitro starch digestibility. *Food Funct.* **2020**, *11*, 3610–3620. [CrossRef]
50. Kaufman, R.C.; Wilson, J.D.; Bean, S.R.; Herald, T.J.; Shi, Y.C. Development of a 96-well plate iodine binding assay for amylose content determination. *Carbohydr. Polym.* **2015**, *115*, 444–447. [CrossRef]
51. Perez-Moral, N.; Plankeele, J.-M.; Domoney, C.; Warren, F.J. Ultra-high performance liquid chromatography-size exclusion chromatography (UPLC-SEC) as an efficient tool for the rapid and highly informative characterisation of biopolymers. *Carbohydr. Polym.* **2018**, *196*, 422–426. [CrossRef]
52. Toutounji, M.R.; Butardo, V.M.; Zou, W.; Farahnaky, A.; Pallas, L.; Oli, P.; Blanchard, C.L. A High-Throughput In Vitro Assay for Screening Rice Starch Digestibility. *Foods* **2019**, *8*, 601. [CrossRef]
53. Cave, R.A.; Seabrook, S.A.; Gidley, M.J.; Gilbert, R.G. Characterization of Starch by Size-Exclusion Chromatography: The Limitations Imposed by Shear Scission. *Biomacromolecules* **2009**, *10*, 2245–2253. [CrossRef]
54. AACC. AACC method 46-30.01, Crude protein—Combustion method. *AACC Approv. Methods Anal.* **2010**. [CrossRef]

foods

MDPI

Article

Foliar Selenate and Zinc Oxide Separately Applied to Two Pea Varieties: Effects on Growth Parameters and Accumulation of Minerals and Macronutrients in Seeds under Field Conditions

Maksymilian Malka [1], Gijs Du Laing [1], Alžbeta Hegedűsová [2] and Torsten Bohn [3,*]

1 Laboratory of Analytical Chemistry and Applied Ecochemistry, Department of Green Chemistry and Technology, Faculty of Bioscience Engineering, Ghent University, Coupure Links 653, 9000 Ghent, Belgium
2 Institute of Horticulture, Faculty of Horticulture and Landscape Engineering, Slovak University of Agriculture in Nitra, Tr. A. Hlinku 2, 94976 Nitra, Slovakia; alzbeta.hegedusova@uniag.sk
3 Nutrition and Health Research Group, Department of Precision Health, Luxembourg Institute of Health, 1 A-B, Rue Thomas Edison, 1445 Strassen, Luxembourg
* Correspondence: torsten.bohn@lih.lu

Abstract: Though selenium (Se) and zinc (Zn) constitute essential nutrients for human health, their deficiencies affect up to 15% and 17% of the global population, respectively. Agronomic biofortification of staple crops with Se/Zn may alleviate these challenges. Pea (*Pisum sativum* L.) is a nutritious legume crop that has great potential for Se/Zn biofortification. Herein, two varieties of pea (Ambassador, Premium) were biofortified via foliar application of sodium selenate (0/50/100 g of Se/ha) or zinc oxide (0/375/750 g of Zn/ha) during the flowering stage under field conditions. While no significant differences were found in Se accumulation between seed varieties upon Se treatments, selenate enhanced the accumulation of Se in the two seed varieties in a dose dependent manner. Selenium concentration was most elevated in seeds of Ambassador exposed to 100 g of Se/ha (3.93 mg/kg DW compared to the control (0.08 mg/kg DW), $p < 0.001$). 375 g of Zn/ha (35.7 mg/kg DW) and 750 g of Zn/ha (35.5 mg/kg DW) significantly and similarly enhanced Zn concentrations compared to the control (31.3 mg/kg DW) in Premium seeds, $p < 0.001$. Zinc oxide also improved accumulations of Fe, Cu, Mn, and Mg in Premium seeds. Se/Zn treatments did not significantly affect growth parameters and accumulations of soluble solids and protein in seeds. Positive and significant ($p < 0.01$) correlations were observed between Zn and Fe, Cu, Mn and Mg levels in Premium seeds, among others. Consuming 33 g/day of pea biofortified with Se at 50 g/ha and 266 g/day of pea biofortified with 375 g of Zn/ha could provide 100% of the RDA (55 μg) for Se and RDA (9.5 mg) for Zn in adults, respectively. These results are relevant for enhancing Se/Zn status in peas by foliar biofortification.

Keywords: crop biofortification; selenium and zinc; trace element deficiency; nutrition; food supply

Citation: Malka, M.; Laing, G.D.; Hegedűsová, A.; Bohn, T. Foliar Selenate and Zinc Oxide Separately Applied to Two Pea Varieties: Effects on Growth Parameters and Accumulation of Minerals and Macronutrients in Seeds under Field Conditions. *Foods* **2023**, *12*, 1286. https://doi.org/10.3390/foods12061286

Academic Editors: Jennifer Ahn-Jarvis and Brittany A. Hazard

Received: 14 February 2023
Revised: 10 March 2023
Accepted: 13 March 2023
Published: 17 March 2023

1. Introduction

Selenium (Se) and zinc (Zn) are indispensable micro-elements in the human body. Selenium (as selenoproteins) and Zn act as cofactors for many important antioxidant enzymes, such as glutathione-peroxidase (GPx) and superoxide-dismutase (SOD), respectively [1,2]. Deficiency of Se and Zn can lead to increased risk of mortality and development of multi-factorial illnesses, both communicable and non-communicable [1,3–6]. However, higher intakes have also been reported to produce adverse health effects, as reviewed previously [7,8].

Selenium and Zn deficiencies affect up to 15% and 17% of the global population, respectively [9,10]. It should be stressed that both deficiency and suboptimal status in micronutrients negatively affect human health. Reports have emphasized widespread below-optimal levels of Se and Zn throughout Europe. This is largely attributable to inadequate concentrations in soil [11], reflecting depletions by agricultural practices and rainfall. Dietary Se/Zn availability is associated to a great extent with the soil Se/Zn level and the bioavailability of Se/Zn from main crops [12–14], though Se/Zn in soil is not homogenous in its distribution and availability to crops [15,16].

Agronomic biofortification is a promising approach that aims to enhance the nutritional value (mostly micronutrients) of staple crops. This strategy relies on optimized application of fertilizers to the soil and/or crop leaves (foliar fertilization), as reviewed previously [17–19]. Foliar fertilization is a very efficient way of Se and Zn plant biofortification [20,21]. In addition, under field conditions, foliar fertilization is less damaging to the environment, preventing accumulation of the fortificant in the soil, leading potentially to contamination [20,22]. The efficacy of foliar application of trace elements depends on a number of factors, namely the physico-chemical specifications of the formulation, the environmental conditions during the time of spraying, or the characteristics of the plant—such as leaf size and metabolism—to which spraying is applied, as summarized previously [23].

Although legumes are staple food crops for billions of people worldwide, their biofortification is still insufficiently used as an approach for alleviating hidden hunger [24,25]. Pea (*Pisum sativum* L.) is a cool season legume harvested all over the world, mainly in temperate regions [26]. The global harvest of dry peas in 2020 was reported as 14.6 million tons, with cultured areas spanning 7.2 million ha (FAOSTAT 2022). Pea is an important human and animal food crop that plays critical roles (as do other pulses) in sustainable agriculture, biodiversity and the environment, as well as in global health and food security [27–29]. Peas are an important source of protein and carbohydrates such as starch, but also provide dietary fibre, minerals and vitamins, in addition to phytochemicals, among others [30,31]. Pea protein is a high-biological-value protein that exhibits health benefits encompassing antioxidant, antihypertensive, and prebiotic-like effects on the gut microbiota [32]. Consumption of peas and their components has been linked to aspects of cardiometabolic as well as gastrointestinal health [30,33].

Consequently, in this investigation we studied the response of two pea varieties (Ambassador, Premium) to foliar application of Se and Zn during the flowering stage under field conditions. Growth parameters, accumulation of selected trace elements (including Se and Zn), macro elements, soluble solids and protein in seeds, as well as their respective correlations were evaluated and compared with previous findings, including our pot experiment [34–36].

2. Materials and Methods

2.1. Chemicals

Zinkuran SC was obtained from Arysta LifeScience Slovakia s.r.o. (Nové Zámky, Slovakia). Sodium selenate was purchased from Alfa Aesar (Karlsruhe, Germany). HNO_3 and H_2O_2 (suprapure quality) were ordered from LGC Standards (Molsheim, France) and Merck/VWR (Leuven, Belgium), respectively.

2.2. Experimental Design and Sample Preparation

A field experiment was carried out in 2014 at the Slovak University of Agriculture in Nitra/Botanical Garden (Nitra, 48.305 N, 18.096 E). The design consisted of four replicates, two varieties of pea (*Pisum sativum* L.) and five treatments, totalling 40 plots. The area of each plot was 1 m^2. Pea was sown at a rate of 85 seeds / m^2 at a seeding depth of 5 cm. To minimize potential Se/Zn cross-contamination, border rows were laid out between each plot. Soil analysis and weather data during the growing season (April–June) in Nitra in 2014 are presented in Table 1. Soil element concentrations were measured using the procedure of Varényiová et al. [37] for total N, and available P, K, Mg, and Ca; the methods of Ducsay et al. [38] and Lindsay and Norvell [39] were used to determine total Se and available Zn, respectively. Seeds of two high-yielding and dark-seeded pea varieties were selected for the investigation. Ambassador (representing late variety/restored hybrid) and Premium (constituting an early variety/open pollinated) were obtained from a farmer from the area of Nitra. Selenium (sodium selenate) and Zn (Zinkuran SC (30% ZnO + 6% chelate)) were administered as part of the following treatments: unaltered control, 50 g (Se1), 100 g (Se2), 375 g (Zn1) and 750 g (Zn2), all per ha. The applied solutions had final concentrations of 0.1 and 0.2 g/L of Se or 0.75 and 1.5 g/L of Zn. Foliar application of Se/Zn was conducted at the flowering stage of the pea plants within dry weather periods using a manual sprayer. Additional fertilizers were not employed. Regular irrigation and phytosanitary measures were applied during the experiment. Incidences of pests/diseases and adverse/toxic impacts of foliar Se/Zn applications on plants were not reported. Seeds were harvested at physiological maturity. The soluble solids concentration was determined in fresh seeds, while the remainder of the seeds was rapidly lyophilized and ground to determine the concentration of trace (Se, Zn, Fe, Cu, Mn, Mo) and macro elements (Ca, Mg, K, Na) and protein.

Table 1. Physical and chemical variables of tested soil, mean monthly temperature (air) and total monthly rainfall (April–June) reported in the experimental area in Nitra in 2014.

pH (H_2O)	pH (KCl)	N	P	K	Ca	Mg	Zn	Se	C_{ox} (%)	Humus (%)	Soil Type
					mg kg^{-1}						
7.55	6.36	13.3	253	285	5630	364	2.47	0.08	1.39	4.01	Gleyic fluvisol
Average air temperature (°C) and total rainfall (mm) in brackets											
	April				May					June	
	12.4 (48.9)				15.2 (57.6)					19.3 (52.5)	

C_{ox}: oxidizable carbon.

2.3. Growth Measures

The number of seeds/pod, and length, perimeter and width of the pod were determined following harvest. Drying of samples was carried out in a drying oven (105 °C) until no more weight changes were noted prior to seed dry matter measurements.

2.4. Measured Soluble Solids and Protein

The level of soluble solids (SSC) was determined with a handheld digital refractometer (DR201-95, A. KRÜSS Optronic, Hamburg, Germany). All samples were measured in duplicates; the average was then expressed as the SSC value (% fresh weight (FW)).

Protein level (in duplicates) was measured following the Dumas technique using the TruSpec CHNS analyser (LECO, Saint Joseph, MI, USA). The values obtained were multiplied by the respective nitrogen conversion factor, 5.4 [40], to derive the amount of protein in the investigated samples.

2.5. Concentration of Minerals

Following Kaulmann et al. [41], 250 mg of freeze-dried pea was mineralised in 7 mL HNO_3 (68%) and 3 mL H_2O_2 (30%). The mineralisation was done in PTFE vials in a microwave furnace (Multiwave Pro, Anton Paar, Graz, Austria), by raising the temperature and pressure to 200 °C and 30 bars, respectively. Afterwards, samples were diluted to 25 mL with ultra-pure water. A reference vegetable (Spinach, NCS ZC 73013, LGC Standards, Molsheim, France) was employed during each mineralisation cycle. Samples were analysed by ICP-MS (Elan DRC-e, Perkin Elmer, Waltham, MA, USA).

2.6. Statistical Analysis

The normality of data distribution (normality plots) and variance equality (box plots) were tested. When needed, the data was logarithmically transformed to better fit a normal distribution. Afterwards, multivariate models were created, with the number of seeds/pod, pod length, pod perimeter, pod width, seed dry matter, and concentrations of soluble solids, protein, Se, Zn, Fe, Cu, Mn, Mo, Ca, Mg, K, and Na in seeds as the dependent observed variables, and pea variety (2 levels) and level of biofortification (5 levels, 2 for Se, 2 for Zn, and controls) as fixed factors. The different levels of biofortification were nested within the biofortificant. Whenever significant results from Fisher-F tests were obtained, all possible group-wise comparisons were conducted (Bonferroni post-hoc tests). When interactions were significant, models were run again by keeping one of the interacting terms constant. p-values < 0.05 (2-sided) were regarded as significant. SPSS (vs. 25.0, IBM, Chicago, IL, USA) was chosen for all statistical investigations.

3. Results

3.1. General Effects

After constructing the multivariate models, combined analysis of variance indicated that treatment (pooled varieties) had a significant impact on the levels of Se, Zn, Fe, Cu, Mn, Mg, and protein in seeds. Variety (pooled treatments) revealed a significant impact on all variables except for the concentration of Se, K, and soluble solids in seeds. The treatment x variety interaction was significant in part (Table 2).

Table 2. Combined ANOVA for the effects of treatment and variety on number of seeds/pod, length, perimeter and width of the pod, seed dry matter, and levels of soluble solids, protein, Se, Zn, Fe, Cu, Mn, Mo, Ca, Mg, K, and Na in seeds.

	Treatment	**Variety**	**Treatment × Variety**
DF	4	1	4
Number of seeds/pod	NS	<0.001	NS
Pod length (cm)	NS	<0.001	NS
Pod perimeter (cm)	NS	<0.001	NS
Pod width (cm)	NS	<0.001	NS
Seed dry matter (%)	NS	<0.001	NS
Soluble solids (% FW)	NS	NS	NS
Protein (% DW)	0.002	<0.001	<0.001
Se (mg/kg DW)	<0.001	NS	NS
Zn (mg/kg DW)	0.002	<0.001	0.002
Fe (mg/kg DW)	<0.001	<0.001	0.011
Cu (mg/kg DW)	0.007	<0.001	<0.001
Mn (mg/kg DW)	0.004	0.004	NS
Mo (mg/kg DW)	NS	<0.001	NS
Ca (mg/kg DW)	NS	<0.001	NS
Mg (mg/kg DW)	0.004	<0.001	0.001
K (mg/kg DW)	NS	NS	NS
Na (mg/kg DW)	NS	<0.001	NS

DF: degrees of freedom; NS: not significant.

3.2. Growth Parameters

Treatment had no significant impact on the number of seeds/pod, length, perimeter and width of the pod, and seed dry matter vs. controls of both varieties. Ambassador had a significantly higher number of seeds/pod, length, perimeter and width of the pod than Premium for all treatments. Ambassador demonstrated slightly but significantly more elevated seed dry matter compared to Premium for all treatments except for Se2 (Figure 1A–E).

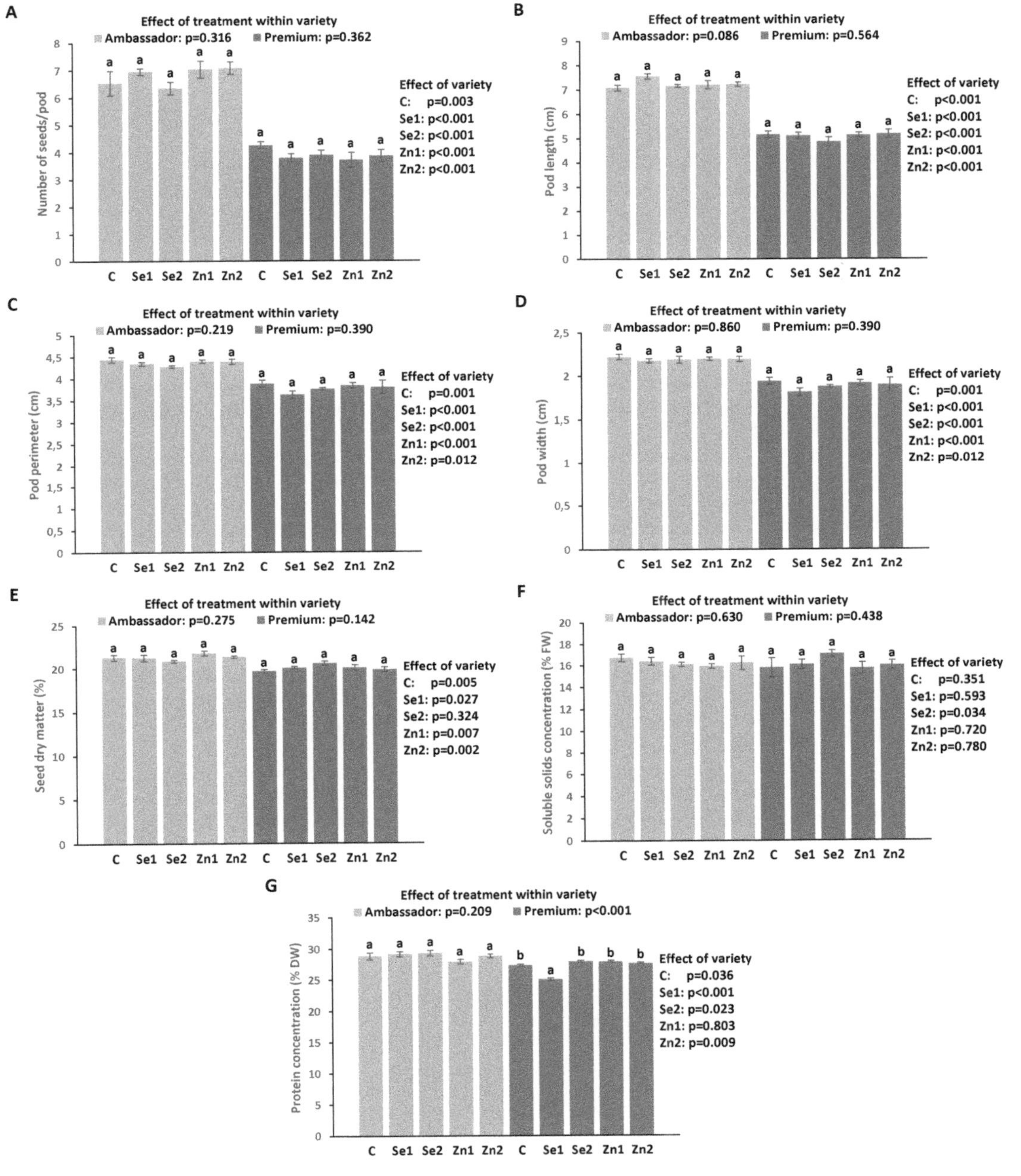

Figure 1. Impact of foliar Se and Zn treatments and variety on the number of seeds/pod (**A**), pod length (**B**), pod perimeter (**C**), pod width (**D**), seed dry matter (**E**), and concentrations of soluble solids (**F**) and protein (**G**) in seeds. C (control): no applied Se and Zn; Se1: 50 g; Se2: 100 g; Zn1: 375 g; Zn2: 750 g; all per ha; mean ± SD; n = 4 replicates. Bars without identical lower letters differed significantly within variety. *p*-values given on the right part of the graph indicate the impact of treatment across varieties (Ambassador compared to Premium).

3.3. Soluble Solid Levels (SSC) in Seeds

Treatment had no significant influence on SSC vs. controls in the two varieties. Premium showed slightly but significantly elevated SSC vs. Ambassador for the Se2 treatment (Figure 1F).

3.4. Protein Levels in Seeds

Biofortification did not significantly impact protein levels vs. the control in Ambassador, while Se1 application significantly reduced protein levels vs. the control in Premium. Ambassador exhibited significantly elevated protein levels compared to Premium for all treatment conditions but not for Zn1 (Figure 1G).

3.5. Concentration of Trace Elements in Seeds

Selenium biofortification significantly improved Se levels vs. controls (in both varieties). The highest concentration of Se was encountered in Ambassador exposed to Se2 vs. the control, followed by Premium exposed to Se2 vs. the control. Differences found in Se levels between Ambassador and Premium for Se1 and Se2 were not significant. Ambassador had slightly though significantly higher levels of Se than Premium for the control (Figure 2A).

Zinc treatments did not significantly affect Zn levels compared to the control in Ambassador, while Zn1 and Zn2 significantly elevated Zn levels vs. the control in Premium. No significant differences were found regarding the concentration of Zn between Ambassador and Premium for Zn1 and Zn2. Ambassador displayed significantly raised Zn concentration vs. Premium for the control (Figure 2B).

The Zn2 treatment significantly improved Fe levels vs. the control in Ambassador, while Zn1 and Zn2 significantly raised Fe levels vs. the control in Premium. Ambassador exhibited significantly elevated levels of Fe relative to Premium for all treatments but not for Zn1 (Figure 2C).

Biofortification had no significant impact on Cu levels vs. the control in Ambassador, while Zn1 and Zn2 significantly raised Cu levels vs. the control in Premium. Ambassador had significantly higher Cu levels than Premium for all biofortification treatments except for Zn1 (Figure 2D).

Treatment failed to significantly impact Mn levels vs. the control in Ambassador, while Zn1 and Zn2 significantly raised Mn levels vs. the control in Premium. No significant differences were found regarding concentrations of Mn between Ambassador and Premium (all treatments, Figure 2E).

Treatment had no significant impact on Mo levels vs. controls in both varieties. Premium displayed significantly greater concentrations of Mo than Ambassador for all treatments (Figure 2F).

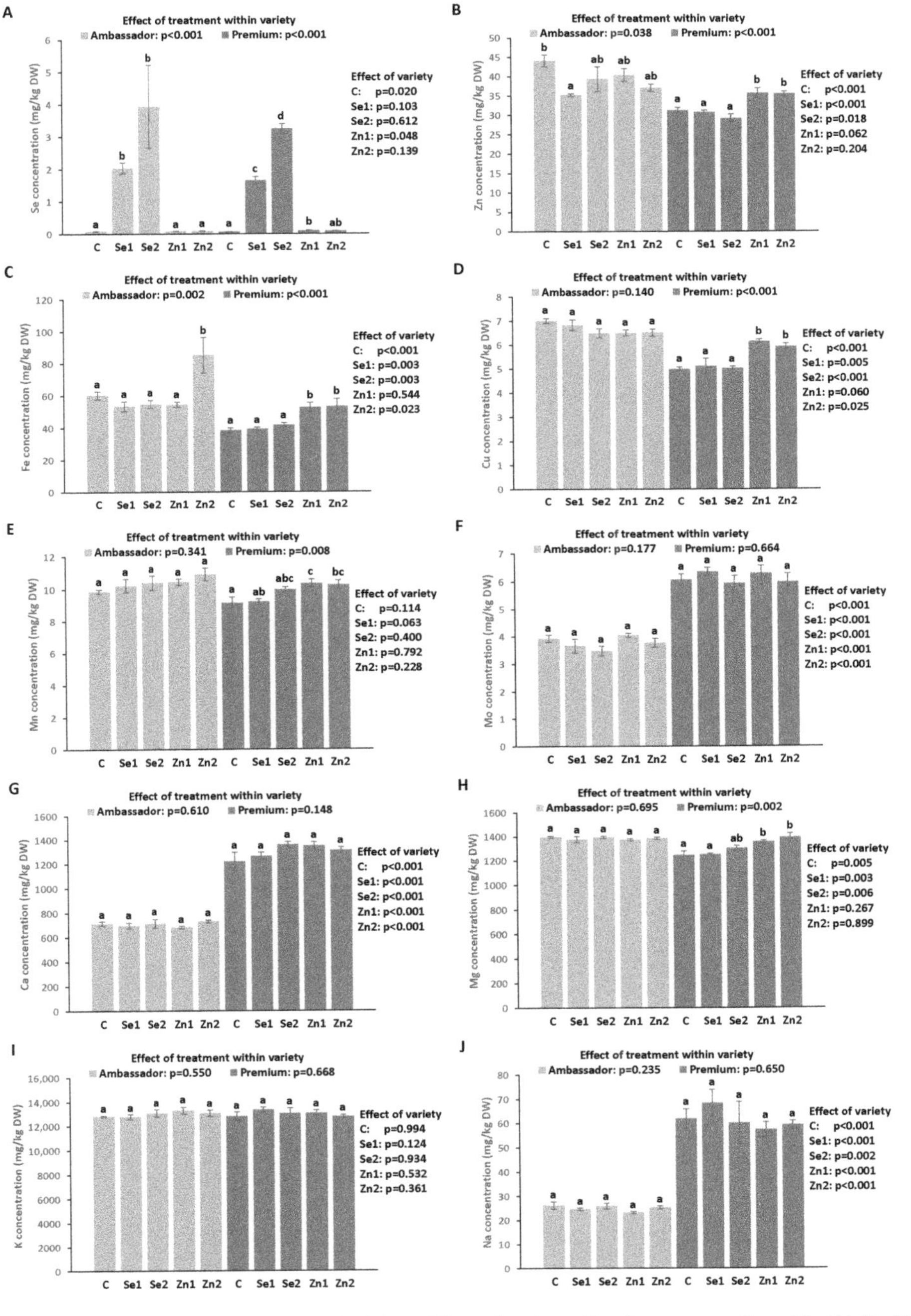

Figure 2. Foliar-applied Se and Zn and variety affect the concentration of Se (**A**), Zn (**B**), Fe (**C**), Cu (**D**), Mn (**E**), Mo (**F**), Ca (**G**), Mg (**H**), K (**I**), and Na (**J**) in seeds. C (control): without Se and Zn; Se1: 50 g; Se2: 100 g; Zn1: 375 g; Zn2: 750 g; all per ha; mean ± SD; n = 4. Bars that do not share an identical lower letter differ significantly within variety. *p*-values on the right part of the graph display the impact of treatment across varieties (Ambassador compared to Premium).

3.6. Concentration of Macro Elements in Seeds

Treatment did not significantly influence Ca levels compared to controls in both varieties. Premium exhibited significantly elevated concentrations of Ca than Ambassador for all treatments (Figure 2G).

Treatment had no significant impact on Mg levels when compared to the control in Ambassador, while Zn1 and Zn2 significantly increased Mg levels vs. the control in Premium. Ambassador had significantly elevated levels of Mg relative to Ambassador for the control, Se1, and Se2 (Figure 2H).

Biofortification did not significantly affect K levels compared to controls in the two varieties. No significant differences were revealed in terms of the concentrations of K between Ambassador and Premium (all treatments, Figure 2I).

The treatment failed to significantly influence Na concentrations compared to controls in both varieties. Premium exhibited significantly elevated levels of Na vs. Ambassador (all treatments, Figure 2J).

3.7. Correlations

For Ambassador, a positive and significant correlation was encountered between the number of seeds/pod and pod length and seed dry matter, between pod length and Mn concentration, between pod circumference and pod width, between Se concentration and pod width, between soluble solids and Zn and Mg concentrations, between Zn and Mo and Ca concentrations, between Mn and Fe, and between Ca and K concentrations, and between Ca and Mg and Na concentrations. A negative and significant correlation was encountered between the number of seeds/pod and pod width, and between Se and Mo concentrations. For Premium, a significant and positive correlation was found between the number of seeds/pod and pod length, pod circumference, pod width and Zn concentration, between pod circumference and pod width, between pod length and protein concentration, between seed dry matter and soluble solid and Se concentrations, between Zn and Fe, Cu, Mn and Mg levels, between Fe and Cu, and between Mg and Mn levels, between Cu and Mn and Mg levels, between Mn and Ca and Mg concentrations, and between K and Na concentrations. A negative and significant correlation was observed between seed dry matter and the number of seeds/pod, pod length and protein concentration, between Ca concentration and pod circumference and pod width, between Na and protein concentrations, between Se and Zn and Cu levels, and between Se concentration and the number of seeds/pod (Figure 3).

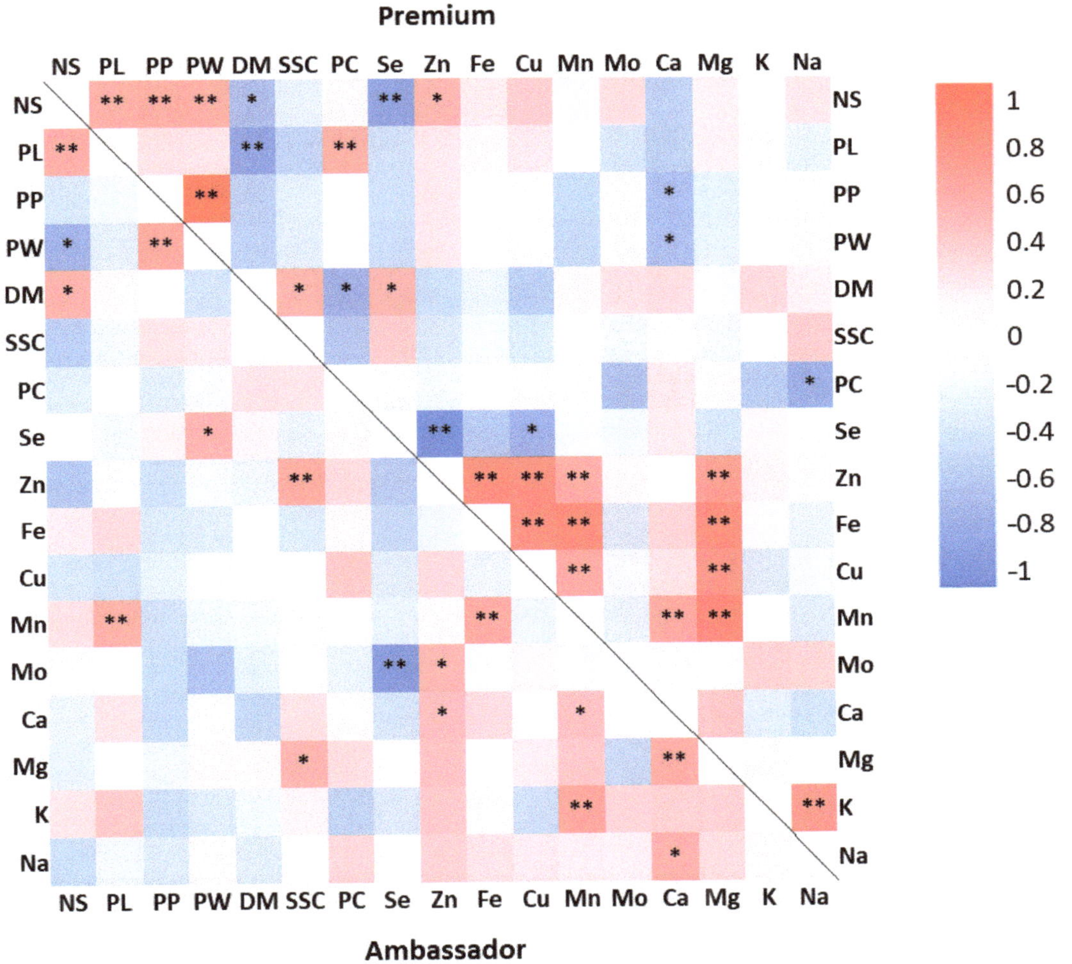

Figure 3. Correlation coefficients (Pearson) between growth parameters and mineral and macronutrient concentrations in seeds determined for the two pea varieties studied (Ambassador, Premium). NS: number of seeds/pod; PL: pod length; PP: pod perimeter; PW: pod width; DM: seed dry matter; SSC: soluble solid concentration; PC: protein concentration. Correlations below and above the black line indicate Ambassador and Premium, respectively. Significance level, * $p < 0.05$, ** $p < 0.01$.

4. Discussion

In this study, we scrutinized the consequences of the foliar application of selenate and zinc oxide on two *Pisum sativum* L. varieties (Ambassador, Premium) during the flowering stage. Parameters related to growth and concentrations of several trace elements (including Se and Zn), macro elements, and macronutrients were determined in seeds. Selenate improved the concentration of Se in both pea varieties (Table 2, Figure 2A). In contrast, zinc oxide enhanced levels of Zn, Fe, Cu, Mn, and Mg in Premium (Table 2, Figure 2). Generally, Se/Zn treatments did not significantly impact growth parameters and concentrations of soluble solids and protein (Table 2, Figure 1).

In this investigation, we have focused on pea, as the crop represents an important staple legume and has been emphasized as a major player to assure food security worldwide, especially due to its high protein concentration. Peas have also been found to be able to accumulate Se and Zn more efficiently compared to cereals, which is likely also related to their high protein concentration [42,43]. The pea varieties included in the study were chosen due to their reported high yield. Foliar application has been described as an efficient method for biofortification, bypassing the effect that soil may have on the bioavailability of examined minerals [34]. While Zn is an indispensable mineral vital for plants [44], the essentiality of Se to higher plants is yet to be established. However, at low concentrations, Se exerts positive impact on plant development and yield, and improves abiotic stress resilience [45]. Regarding the forms of application, selenate was reported to exhibit a high efficiency for foliar uptake [46], while zinc oxide was chosen as recommended by the agro-industry. Our recent findings showed a positive effect of foliar-applied selenate and zinc oxide on the aggregation of phenolics in pea seeds [36]. In addition, upon these applications, total condensed tannins did not correlate with a lower accretion of selected minerals, macronutrients and plant bioactive constituents in pea seeds, which is encouraging in light of agronomic biofortification [34], as tannins have been reported to potentially hamper the nutritional value of many food items, including legumes, by interfering with the digestion and bioavailability of nutrients, including especially divalent minerals [47].

Selenate and zinc oxide did not specifically influence growth parameters (Table 2, Figure 1A–E), which is in line with our pot study (other than in part for the number of seeds/pod upon Zn treatment) [35]. Similar findings were also found by earlier foliar biofortification of peas with selenate and zinc sulphate [43], as well as with selenate and selenite [48]. However, Pandey et al. [49] reported a positive impact of foliar-applied zinc sulphate on the yield parameters of field pea.

Selenate and zinc oxide did not produce beneficial changes on the levels of soluble solids in pea seeds (Table 2, Figure 1F), which is consistent with our pot study [34]. However, previous studies did report increased concentrations of total sugars in seeds of pea [49], black gram [50], and common bean [51] following zinc sulphate application. Soluble solids are mostly simple carbohydrates that are needed for seed development [49]. Carbohydrate metabolism may be positively or negatively affected by selenium, depending on its employed concentration and form, as well as the plant developmental stage, as emphasized previously [52]. The negative impact of Se applications on soluble solids in pea seeds may be related to the lowered content of chlorophyll in seeds [34]. Regarding Zn, its lower status was linked to reduced photosynthesis and activity of sucrose synthase, which consequently impaired sugar synthesis in seeds, as emphasized earlier [49].

Selenate and zinc oxide did not improve protein levels in pea seeds (Table 2, Figure 1G). In contrast, our pot study showed that these applications enhanced protein concentration in Premium seeds [34]. Other research findings have revealed both beneficial and non-significant influences of foliar-applied selenate and/or zinc sulphate on protein levels in pea seeds [42,43,48,53]. The present study further found that Se and Zn concentrations were not significantly correlated with protein concentrations in Ambassador and Premium seeds (Figure 3). However, under pot experimental conditions, the concentration of Zn exhibited a positive and significant correlation with protein levels in both seed varieties [34]. Zinc may influence protein synthesis according to its involvement in DNA/RNA metabolism, chromatin condensation, and gene expression [54]. As previously reported, with respect to Se, its major species in pea seeds following foliar application of selenate was selenomethionine, with relevant antioxidant activity and possible benefits to human and animal health [35,55].

The detected concentration of total Se in the experimental soil (Table 1) is considered deficient according to Gupta and Gupta [56]. Ambassador variety in general showed higher concentrations of Se in seeds than Premium variety upon selenate applications (Figure 2A). In contrast, our pot study demonstrated greater accumulation of Se in the seeds of Premium rather than Ambassador variety following these applications [35]. The positive impact of foliar-applied selenate on the level of Se in pea seeds was also reported by other research [42,43,48,55]. The positive linear relationship between selenate application dose and seed Se level (Table S1) is in accordance with earlier research on pea [35,42,48].

The concentration of Zn determined in the experimental soil (Table 1) is considered to be low, according to previous reports [57,58]. Zinc oxide applications improved Zn concentration only in Premium seeds (Figure 2B). However, the correlation between Zn application dose and Zn concentration in Premium seeds was not significant (Table S1). In contrast, our pot study demonstrated that zinc oxide applications had no positive influence on Zn concentrations in Ambassador and Premium seeds [35]. However, Pandey et al. [49] revealed that foliar application of zinc sulphate enhanced Zn concentration in pea seeds. These findings suggest that zinc sulphate should be used in further investigations on foliar Zn biofortification of peas. However, other and novel Zn forms may also deserve further research, such as ZnEDTA and a Zn–glycine complex, which demonstrated lower phytotoxicity than zinc sulphate [59], allowing applications at a wider dose range.

Selenate, unlike zinc oxide, improved, although in part and not consistently, the concentrations of Fe, Cu, and Mn, while both Se and Zn applications had no favourable effect on Mo levels in pea seeds (Table 2, Figure 2C–F). In contrast, our pot study showed that these applications did not enhance concentrations of Fe, Cu, and Mn in pea seeds [36]. Likewise, other research has demonstrated that foliar-applied selenate and zinc sulphate did not positively influence the level of Fe in the seeds of pea [42,43,53]. Similarly, Kayan et al. [60] reported a non-significant influence of foliar-applied zinc chelate/sulphate on Fe levels in chickpea seeds.

Selenate and zinc oxide failed to significantly impact macroelement concentrations except for that of Mg in Premium seeds upon Zn applications (Table 2, Figure 2G–J). In contrast, our pot study revealed that these applications did not significantly affect the concentrations of Ca, Mg, K, and Na in pea seeds [34]. Likewise, previous research demonstrated that foliar-applied selenate and zinc sulphate had no beneficial effects on levels of Ca and Mg in pea seeds [42,43,53].

The significant relationships observed between minerals in pea seeds (Figure 3) are consistent in part with our earlier findings reported under pot experimental conditions [34–36], namely the positive correlations between Zn and Fe, Cu, Mn, and Mg, as well as the negative one between Se and Zn. Several aspects regarding interactions between accumulation of trace elements were discussed by our previous work [36]. A number of factors may impact the association between minerals in plants, including plant species/genotype, fertilizer dose, form and way of application, conditions of cultivation, soil properties, as well as antagonism and/or synergism between various elements, as emphasized by Malka et al. [34].

Finally, consuming 33 g/day of pea biofortified with Se at 50 g/ha and 266 g/day of pea bio-fortified with 375 g of Zn/ha could provide 100% of the RDA (55 µg) for Se and RDA (9.5 mg) for Zn in adults, respectively (Table 3).

Table 3. Selenium/Zn intake and fraction (%) of recommended dietary allowance for Se/Zn (% RDA) obtained from 100 g of pea seeds.

Se				
Variety	**Se Treatment (g of Se/ha)**	**Se Intake from 100 g (µg/day)**	**% RDA from 100 g (USDA *)**	**% RDA from 100 g (EFSA *)**
Ambassador				
	0	8	15	12
	50	203	369	290
	100	393	715	562
Premium				
	0	7	12	10
	50	165	300	235
	100	325	591	464

Table 3. *Cont.*

Zn				
Variety	**Zn Treatment (g of Zn/ha)**	**Zn Intake from 100 g (mg/day)**	**% RDA from 100 g (USDA *)**	**% RDA from 100 g (EFSA *)**
Ambassador				
	0	4.40	46	38
	375	4.04	42	35
	750	3.71	39	32
Premium				
	0	3.13	33	27
	375	3.57	38	31
	750	3.55	37	31

* 55 and 70 μg are the RDA and AI (adequate intake) set by the USDA and EFSA, respectively, for Se; * 9.5 and 11.5 mg are the RDA and AI, representing averages for males and females for Zn, according to USDA and EFSA, respectively.

5. Conclusions

Taken together, the present investigation demonstrated that selenate and zinc oxide had no distinct adverse impact on the growth parameters of pea, in accordance with an absence of toxic effects. Zinc oxide improved seed Zn accumulation only in one pea variety, and as similar effects were encountered in our pot experiment, more bioavailable alternative forms of Zn, namely zinc sulphate, should be employed in future investigations of peas. Contrarily, selenate considerably increased seed Se accretion in both varieties in a dose-dependent manner. Small portions of pea biofortified with 50 g of Se/ha and 375 g of Zn/ha would cover daily intake recommendations for Se and Zn, respectively; however, lower selenate doses could be envisioned in further research. In addition, the use of nanoparticles for foliar applications has been receiving some attention, but no results on peas are currently available. Foliar Se biofortification of pea could be an encouraging approach to improve human micronutrient supply.

Supplementary Materials: The ensuing supporting data can be downloaded at: https://www.mdpi.com/article/10.3390/foods12061286/s1, Table S1: Correlation analysis between Se/Zn application dose and seed Se/Zn concentration in the studied pea varieties.

Author Contributions: M.M. planned and conducted the experiments, and also drafted the first version of the manuscript. T.B. was implicated in the analyses, statistical evaluation, and further refinement of the manuscript. G.D.L. participated in the design and supervision of the study, and in the final manuscript preparation. A.H. organized, designed, and supervised the experiments in Nitra. All authors have read and agreed to the published version of the manuscript.

Funding: Support for this investigation was received from the Ministry of Education, Science, Research and Sport of the Slovak Republic (grant: VEGA 1/0105/14) and the Luxembourg Institute of Science and Technology.

Institutional Review Board Statement: Not applicable.

Informed Consent Statement: Not applicable.

Data Availability Statement: No new data were created or analyzed in this study. Data sharing is not applicable to this article.

Acknowledgments: We are thankful for the staff of the Slovak University of Agriculture in Nitra for aiding in the set-up of the experiments.

Conflicts of Interest: The authors have no conflicts of interest to declare.

References

1. Barchielli, G.; Capperucci, A.; Tanini, D. The Role of Selenium in Pathologies: An Updated Review. *Antioxidants* **2022**, *11*, 251. [CrossRef]
2. Marreiro, D.D.N.; Cruz, K.J.C.; Morais, J.B.S.; Beserra, J.B.; Severo, J.S.; de Oliveira, A.R.S. Zinc and Oxidative Stress: Current Mechanisms. *Antioxidants* **2017**, *6*, 24. [CrossRef] [PubMed]

3. Kaur, K.; Gupta, R.; Saraf, S.A.; Saraf, S.K. Zinc: The Metal of Life. *Compr. Rev. Food Sci. Food Saf.* **2014**, *13*, 358–376. [CrossRef] [PubMed]
4. Rayman, M.P. Selenium and Human Health. *Lancet* **2012**, *379*, 1256–1268. [CrossRef] [PubMed]
5. Alexander, J.; Tinkov, A.; Strand, T.A.; Alehagen, U.; Skalny, A.; Aaseth, J. Early Nutritional Interventions with Zinc, Selenium and Vitamin D for Raising Anti-Viral Resistance Against Progressive COVID-19. *Nutrients* **2020**, *12*, 2358. [CrossRef]
6. Du Laing, G.; Petrovic, M.; Lachat, C.; De Boevre, M.; Klingenberg, G.J.; Sun, Q.; De Saeger, S.; De Clercq, J.; Ide, L.; Vandekerckhove, L.; et al. Course and Survival of COVID-19 Patients with Comorbidities in Relation to the Trace Element Status at Hospital Admission. *Nutrients* **2021**, *13*, 3304. [CrossRef] [PubMed]
7. Rayman, M.P. Selenium Intake, Status, and Health: A Complex Relationship. *Hormones* **2020**, *19*, 9–14. [CrossRef]
8. Agnew, U.M.; Slesinger, T.L. Zinc Toxicity. In *StatPearls*; StatPearls Publishing: Treasure Island, FL, USA, 2022.
9. Shreenath, A.P.; Ameer, M.A.; Dooley, J. Selenium Deficiency. In *StatPearls*; StatPearls Publishing: Treasure Island, FL, USA, 2022.
10. Kumssa, D.B.; Joy, E.J.M.; Ander, E.L.; Watts, M.J.; Young, S.D.; Walker, S.; Broadley, M.R. Dietary Calcium and Zinc Deficiency Risks Are Decreasing but Remain Prevalent. *Sci. Rep.* **2015**, *5*, 10974. [CrossRef]
11. Tóth, G.; Hermann, T.; Szatmári, G.; Pásztor, L. Maps of Heavy Metals in the Soils of the European Union and Proposed Priority Areas for Detailed Assessment. *Sci. Total Environ.* **2016**, *565*, 1054–1062. [CrossRef]
12. Alloway, B.J. *Zinc in Soils and Crop Nutrition*, 2nd ed.; International Zinc Association: Brussels, Belgium; International Fertilizer Industry Association: Paris, France, 2008.
13. Chilimba, A.D.C.; Young, S.D.; Black, C.R.; Rogerson, K.B.; Ander, E.L.; Watts, M.J.; Lammel, J.; Broadley, M.R. Maize Grain and Soil Surveys Reveal Suboptimal Dietary Selenium Intake Is Widespread in Malawi. *Sci. Rep.* **2011**, *1*, 72. [CrossRef]
14. Winkel, L.H.; Vriens, B.; Jones, G.D.; Schneider, L.S.; Pilon-Smits, E.; Bañuelos, G.S. Selenium Cycling across Soil-Plant-Atmosphere Interfaces: A Critical Review. *Nutrients* **2015**, *7*, 4199–4239. [CrossRef] [PubMed]
15. Sadeghzadeh, B. A Review of Zinc Nutrition and Plant Breeding. *J. Soil Sci. Plant Nutr.* **2013**, *13*, 905–927. [CrossRef]
16. Jones, G.D.; Droz, B.; Greve, P.; Gottschalk, P.; Poffet, D.; McGrath, S.P.; Seneviratne, S.I.; Smith, P.; Winkel, L.H. Selenium Deficiency Risk Predicted to Increase under Future Climate Change. *Proc. Natl. Acad. Sci. USA* **2017**, *114*, 2848–2853. [CrossRef]
17. Szerement, J.; Szatanik-Kloc, A.; Mokrzycki, J.; Mierzwa-Hersztek, M. Agronomic Biofortification with Se, Zn, and Fe: An Effective Strategy to Enhance Crop Nutritional Quality and Stress Defense—A Review. *J. Soil Sci. Plant Nutr.* **2022**, *22*, 1129–1159. [CrossRef]
18. Cakmak, I.; Kutman, U.B. Agronomic Biofortification of Cereals with Zinc: A Review. *Eur. J. Soil Sci.* **2018**, *69*, 172–180. [CrossRef]
19. de Valença, A.W.; Bake, A.; Brouwer, I.D.; Giller, K.E. Agronomic Biofortification of Crops to Fight Hidden Hunger in Sub-Saharan Africa. *Glob. Food Sec.* **2017**, *12*, 8–14. [CrossRef]
20. Delaqua, D.; Carnier, R.; Berton, R.S.; Corbi, F.C.A.; Coscione, A.R. Increase of Selenium Concentration in Wheat Grains through Foliar Application of Sodium Selenate. *J. Food Compos. Anal.* **2021**, *99*, 103886. [CrossRef]
21. Sattar, A.; Wang, X.; Ul-Allah, S.; Sher, A.; Ijaz, M.; Irfan, M.; Abbas, T.; Hussain, S.; Nawaz, F.; Al-Hashimi, A.; et al. Foliar Application of Zinc Improves Morpho-Physiological and Antioxidant Defense Mechanisms, and Agronomic Grain Biofortification of Wheat (*Triticum aestivum* L.) under Water Stress. *Saudi J. Biol. Sci.* **2021**, *29*, 1699–1706. [CrossRef]
22. Lyons, G. Biofortification of Cereals with Foliar Selenium and Iodine Could Reduce Hypothyroidism. *Front. Plant Sci.* **2018**, *9*, 730. [CrossRef]
23. Fernández, V.; Brown, P.H. From Plant Surface to Plant Metabolism: The Uncertain Fate of Foliar-Applied Nutrients. *Front. Plant Sci.* **2013**, *4*, 289. [CrossRef]
24. Kumar, S.; Pandey, G. Biofortification of Pulses and Legumes to Enhance Nutrition. *Heliyon* **2020**, *6*, e03682. [CrossRef]
25. Rehman, H.M.; Cooper, J.W.; Lam, H.M.; Yang, S.H. Legume Biofortification Is an Underexploited Strategy for Combatting Hidden Hunger. *Plant. Cell Environ.* **2018**, *42*, 52–70. [CrossRef] [PubMed]
26. Rubiales, D.; Ambrose, M.J.; Domoney, C.; Burstin, J. Pea. In *Genetics, Genomics and Breeding of Cool Season Grain Legumes*; CRC Press: Boca Raton, FL, USA, 2011; pp. 1–49.
27. Sahruzaini, N.A.; Rejab, N.A.; Harikrishna, J.A.; Khairul Ikram, N.K.; Ismail, I.; Kugan, H.M.; Cheng, A. Pulse Crop Genetics for a Sustainable Future: Where We Are Now and Where We Should Be Heading. *Front. Plant Sci.* **2020**, *11*, 531. [CrossRef] [PubMed]
28. Ferreira, H.; Pinto, E.; Vasconcelos, M.W. Legumes as a Cornerstone of the Transition Toward More Sustainable Agri-Food Systems and Diets in Europe. *Front. Sustain. Food Syst.* **2021**, *5*, 694121. [CrossRef]
29. Powers, S.E.; Thavarajah, D. Checking Agriculture's Pulse: Field Pea (*Pisum sativum* L.), Sustainability, and Phosphorus Use Efficiency. *Front. Plant Sci.* **2019**, *10*, 1489. [CrossRef]
30. Dahl, W.J.; Foster, L.M.; Tyler, R.T. Review of the Health Benefits of Peas (*Pisum sativum* L.). *Br. J. Nutr.* **2012**, *108*, S3–S10. [CrossRef]
31. Singh, B.; Singh, J.P.; Kaur, A.; Singh, N. Phenolic Composition and Antioxidant Potential of Grain Legume Seeds: A Review. *Food Res. Int.* **2017**, *101*, 1–16. [CrossRef] [PubMed]
32. Ge, J.; Sun, C.X.; Corke, H.; Gul, K.; Gan, R.Y.; Fang, Y. The Health Benefits, Functional Properties, Modifications, and Applications of Pea (*Pisum sativum* L.) Protein: Current Status, Challenges, and Perspectives. *Compr. Rev. Food Sci. Food Saf.* **2020**, *19*, 1835–1876. [CrossRef] [PubMed]
33. Kumari, T.; Deka, S.C. Potential Health Benefits of Garden Pea Seeds and Pods: A Review. *Legum. Sci.* **2021**, *3*, e82. [CrossRef]

34. Malka, M.; Du Laing, G.; Bohn, T. Separate Effects of Foliar Applied Selenate and Zinc Oxide on the Accumulation of Macrominerals, Macronutrients and Bioactive Compounds in Two Pea (*Pisum sativum* L.) Seed Varieties. *Plants* **2022**, *11*, 2009. [CrossRef]
35. Malka, M.; Du Laing, G.; Li, J.; Bohn, T. Separate Foliar Sodium Selenate and Zinc Oxide Application Enhances Se but Not Zn Accumulation in Pea (*Pisum sativum* L.) Seeds. *Front. Plant Sci.* **2022**, *13*, 968324. [CrossRef] [PubMed]
36. Malka, M.; Du Laing, G.; Kurešová, G.; Hegedűsová, A.; Bohn, T. Enhanced Accumulation of Phenolics in Pea (*Pisum sativum* L.) Seeds upon Foliar Application of Selenate or Zinc Oxide. *Front. Nutr.* **2023**, *10*, 1083253. [CrossRef]
37. Varényiová, M.; Ducsay, L.; Ryant, P. Sulphur Nutrition and Its Effect on Yield and Oil Content of Oilseed Rape (*Brassica napus* L.). *Acta Univ. Agric. Silvic. Mendelianae Brun.* **2017**, *65*, 555–562. [CrossRef]
38. Ducsay, L.; Ložek, O.; Varga, L. The Influence of Selenium Soil Application on Its Content in Spring Wheat. *Plant Soil Environ.* **2009**, *55*, 80–84. [CrossRef]
39. Lindsay, W.L.; Norvell, W.A. Development of a DTPA Soil Test for Zinc, Iron, Manganese, and Copper. *Soil Sci. Soc. Am. J.* **1978**, *42*, 421–428. [CrossRef]
40. Mariotti, F.; Tomé, D.; Mirand, P.P. Converting Nitrogen into Protein-Beyond 6.25 and Jones' Factors. *Crit. Rev. Food Sci. Nutr.* **2008**, *48*, 177–184. [CrossRef]
41. Kaulmann, A.; Jonville, M.C.; Schneider, Y.J.; Hoffmann, L.; Bohn, T. Carotenoids, Polyphenols and Micronutrient Profiles of Brassica Oleraceae and Plum Varieties and Their Contribution to Measures of Total Antioxidant Capacity. *Food Chem.* **2014**, *155*, 240–250. [CrossRef]
42. Poblaciones, M.J.; Rengel, Z. The Effect of Processing on Pisum Sativum L. Biofortified with Sodium Selenate. *J. Plant Nutr. Soil Sci.* **2018**, *181*, 932–937. [CrossRef]
43. Poblaciones, M.J.; Rengel, Z. Combined Foliar Selenium and Zinc Biofortification in Field Pea (*Pisum sativum*): Accumulation and Bioavailability in Raw and Cooked Grains. *Crop Pasture Sci.* **2017**, *68*, 265–271. [CrossRef]
44. Hacisalihoglu, G. Zinc (Zn): The Last Nutrient in the Alphabet and Shedding Light on Zn Efficiency for the Future of Crop Production under Suboptimal Zn. *Plants* **2020**, *9*, 1471. [CrossRef]
45. Hasanuzzaman, M.; Bhuyan, M.B.; Raza, A.; Hawrylak-Nowak, B.; Matraszek-Gawron, R.; Al Mahmud, J.; Nahar, K.; Fujita, M. Selenium in Plants: Boon or Bane? *Environ. Exp. Bot.* **2020**, *178*, 104170. [CrossRef]
46. Ros, G.H.; van Rotterdam, A.M.D.; Bussink, D.W.; Bindraban, P.S. Selenium Fertilization Strategies for Bio-Fortification of Food: An Agro-Ecosystem Approach. *Plant Soil* **2016**, *404*, 99–112. [CrossRef]
47. Grela, E.R.; Samolińska, W.; Kiczorowska, B.; Klebaniuk, R.; Kiczorowski, P. Content of Minerals and Fatty Acids and Their Correlation with Phytochemical Compounds and Antioxidant Activity of Leguminous Seeds. *Biol. Trace Elem. Res.* **2017**, *180*, 338–348. [CrossRef] [PubMed]
48. Poblaciones, M.J.; Rodrigo, S.; Santamaría, O. Evaluation of the Potential of Peas (*Pisum sativum* L.) to Be Used in Selenium Biofortification Programs under Mediterranean Conditions. *Biol. Trace Elem. Res.* **2013**, *151*, 132–137. [CrossRef] [PubMed]
49. Pandey, N.; Gupta, B.; Pathak, G.C. Enhanced Yield and Nutritional Enrichment of Seeds of Pisum Sativum L. through Foliar Application of Zinc. *Sci. Hortic. (Amst.)* **2013**, *164*, 474–483. [CrossRef]
50. Pandey, N.; Gupta, B.; Pathak, G.C. Foliar Application of Zn at Flowering Stage Improves Plant's Performance, Yield and Yield Attributes of Black Gram. *Indian J. Exp. Biol.* **2013**, *51*, 548–555.
51. Kachinski, W.D.; Ávila, F.W.; dos Reis, A.R.; Muller, M.M.L.; Mendes, M.C.; Petranski, P.H. Agronomic Biofortification Increases Concentrations of Zinc and Storage Proteins in Common Bean (*Phaseolus vulgaris* L.) Grains. *Food Res. Int.* **2022**, *155*, 111105. [CrossRef]
52. Lara, T.S.; de Lima Lessa, J.H.; de Souza, K.R.D.; Corguinha, A.P.B.; Martins, F.A.D.; Lopes, G.; Guilherme, L.R.G. Selenium Biofortification of Wheat Grain via Foliar Application and Its Effect on Plant Metabolism. *J. Food Compos. Anal.* **2019**, *81*, 10–18. [CrossRef]
53. Poblaciones, M.J.; Rengel, Z. Soil and Foliar Zinc Biofortification in Field Pea (*Pisum sativum* L.): Grain Accumulation and Bioavailability in Raw and Cooked Grains. *Food Chem.* **2016**, *212*, 427–433. [CrossRef]
54. Moreira, A.; Moraes, L.A.C.; Reis, A.R. The Molecular Genetics of Zinc Uptake and Utilization Efficiency in Crop Plants. In *Plant Micronutrient Use Efficiency*; Academic Press: Cambridge, MA, USA, 2018; pp. 87–108. ISBN 9780128122433.
55. Smrkolj, P.; Germ, M.; Kreft, I.; Stibilj, V. Respiratory Potential and Se Compounds in Pea (*Pisum sativum* L.) Plants Grown from Se-Enriched Seeds. *J. Exp. Bot.* **2006**, *57*, 3595–3600. [CrossRef]
56. Gupta, U.C.; Gupta, S.C. Selenium in Soils and Crops, Its Deficiencies in Livestock and Humans: Implications for Management. *Commun. Soil Sci. Plant Anal.* **2000**, *31*, 1791–1807. [CrossRef]
57. Noulas, C.; Tziouvalekas, M.; Karyotis, T. Zinc in Soils, Water and Food Crops. *J. Trace Elem. Med. Biol.* **2018**, *49*, 252–260. [CrossRef] [PubMed]
58. De Groote, H.; Tessema, M.; Gameda, S.; Gunaratna, N.S. Soil Zinc, Serum Zinc, and the Potential for Agronomic Biofortification to Reduce Human Zinc Deficiency in Ethiopia. *Sci. Rep.* **2021**, *11*, 8770. [CrossRef] [PubMed]

59. Xu, M.; Du, L.; Liu, M.; Zhou, J.; Pan, W.; Fu, H.; Zhang, X.; Ma, Q.; Wu, L. Glycine-Chelated Zinc Rather than Glycine-Mixed Zinc Has Lower Foliar Phytotoxicity than Zinc Sulfate and Enhances Zinc Biofortification in Waxy Corn. *Food Chem.* **2022**, *370*, 131031. [CrossRef] [PubMed]
60. Kayan, N.; Gulmezoglu, N.; Kaya, M.D. The Optimum Foliar Zinc Source and Level for Improving Zn Content in Seed of Chickpea. *Legum. Res.* **2015**, *38*, 826–831. [CrossRef]

foods

MDPI

Article

Dietary Fibre Impacts the Texture of Cooked Whole Grain Rice

Siriluck Wattanavanitchakorn [1], Rungtiva Wansuksri [2], Ekawat Chaichoompu [1,3], Wintai Kamolsukyeunyong [4] and Apichart Vanavichit [1,*]

1 Rice Science Center, Kasetsart University, Kamphangsaen, Nakhon Pathom 73140, Thailand
2 Cassava and Starch Technology Research Team, National Center for Genetic Engineering and Biotechnology, National Science and Technology Development Agency, Khlong Nueng 12120, Thailand
3 Interdisciplinary Graduate Program in Genetic Engineering and Bioinformatics, Kasetsart University, Chatuchak, Bangkok 10900, Thailand
4 Innovative Plant Biotechnology and Precision Agriculture Research Team, National Center for Genetic Engineering and Biotechnology (BIOTEC), 113 Thailand Science Park, Phahonyothin Road, Khlong Nueng 12120, Thailand
* Correspondence: vanavichit@gmail.com; Tel.: +66-81-527-4070

Abstract: Consumers' general preference for white rice over whole grain rice stems from the hardness and low palatability of cooked whole grain rice; however, strong links have been found between consuming a large amount of white rice, leading a sedentary lifestyle, and acquiring type 2 diabetes. This led us to formulate a new breeding goal to improve the softness and palatability of whole grain rice while promoting its nutritional value. In this study, the association between dietary fibre profiles (using an enzymatic method combined with high-performance liquid chromatography) and textural properties of whole grain rice (using a texture analyser) was observed. The results showed that a variation in the ratio of soluble dietary fibre (SDF) and insoluble dietary fibre (IDF) influenced the textural characteristics of cooked whole grain rice; found a strong association between SDF to IDF ratio and hardness ($r = -0.74$, $p < 0.01$) or gumminess ($r = -0.69$, $p < 0.01$) of cooked whole grain rice, and demonstrated that the SDF to IDF ratio was also moderately correlated with cohesiveness ($r = -0.45$, $p < 0.05$), chewiness ($r = -0.55$, $p < 0.01$), and adhesiveness ($r = 0.45$, $p < 0.05$) of cooked whole grain rice. It is suggested that the SDF to IDF ratio can be used as a biomarker for breeding soft and highly palatable whole grain rice of cultivated tropical indica rice to achieve consumer well-being. Lastly, a simple modified method from the alkaline disintegration test was developed for high-throughput screening of dietary fibre profiles in the whole grain indica rice samples.

Keywords: whole grain rice; rice bran; soft-texture rice; dietary fibre profiles; biomarker

Citation: Wattanavanitchakorn, S.; Wansuksri, R.; Chaichoompu, E.; Kamolsukyeunyong, W.; Vanavichit, A. Dietary Fibre Impacts the Texture of Cooked Whole Grain Rice. *Foods* **2023**, *12*, 899. https://doi.org/10.3390/foods12040899

Academic Editors: Jennifer Ahn-Jarvis and Brittany A. Hazard

Received: 20 December 2022
Revised: 12 February 2023
Accepted: 16 February 2023
Published: 20 February 2023

1. Introduction

Rice (*Oryza sativa*) is a major staple food in most Asian countries. Consumers increasingly prefer white rice for its features such as texture, palatability, and appearance [1]. A few reports have associated lower consumer acceptability with an increase in the hardness of rice [2–4]. Additionally, statistical estimation reveals that white rice consumption accounts for about 85% of the total global rice consumption [5]. However, a relationship might exist between the consumption of white rice and the development of non-communicable diseases (NCDs). Epidemiological studies in China [6] and Japan [7] have revealed that high consumption of white rice increases the risk of developing type 2 diabetes (T2D) (78% and 65%, respectively), which is consistent with the risk in the Caucasian population in the United States [8]. Furthermore, a meta-analysis and systematic review concluded that the relative risk for T2D was 1.55 for the Asian population compared to 1.12 for the Western population [9]. The soft texture and the absence of bran layer in white rice may result in faster digestion and higher glycaemic index (GI) [10–12].

Whole grain rice or brown rice consists of a bran layer (6–7% of its total weight), germ (2–3%), and starchy endosperm (90–91%); if this whole grain rice is further milled to remove the bran layer and germ, it is termed 'white rice' or 'milled rice' [13]. These by-products, lost during the milling process, have high amounts of macro- and micronutrients, such as protein, lipids, dietary fibre, vitamins, minerals, and phytochemicals (e.g., phenolic acids, flavonoids, anthocyanin, tocopherols, γ-oryzanol, and phytic acid), which are all health-promoting components [14–17]. Current cohort studies and systematic reviews have revealed that a higher intake of whole grain rice is associated with a lower risk of NCDs, such as T2D, cardiovascular diseases (CVDs), and cancers [15,18–23]. This effect is partly due to the high amount of bioactive compounds in rice bran and germ, which have remarkable biological activities, such as anti-oxidant, anti-diabetic, anti-obesity, cholesterol-lowering, anti-cancer, and anti-inflammatory activities [15,22–29].

Dietary fibre, as defined by the American Association of Cereal Chemists (AACC, 2000) [30], refers to the edible parts of a plant or analogous carbohydrates that are resistant to digestion and absorption by the small intestine with complete or partial fermentation in the large intestine in humans. It is further classified into soluble and insoluble types. Soluble dietary fibre (SDF) easily dissolves in water and gastrointestinal fluids, then transforms into a gel-like substance, resulting in the blockage of digestion and absorption of fat and carbohydrate. This causes a reduction in blood cholesterol and sugar levels. Conversely, insoluble dietary fibre (IDF) does not dissolve in water, but absorbs fluid and increases the faecal bulk. This reduces the transit time of food in the digestive tract, thereby preventing gastrointestinal blockage and constipation, which are causes of colorectal cancer [31–33]. Some studies have demonstrated the influence of dietary fibre content in increasing the glycaemic index (GI) value of milled rice [10,34–37], suggesting that the bran layer serves as a physical barrier, which leads to a block in water absorption, inhibition of starch granule swelling during thermal processing, and a decrease in enzyme accessibility [5]. Fibre does not have a GI value, and the addition of fibre in a meal also lowers the GI value of a carbohydrate-rich diet [38]. A growing body of evidence indicates that the regular consumption of a fibre-rich diet with whole grain rice prevents the risk of T2D [19,21,39]. The consumption of dietary fibre plays a vital role in maintaining healthy gut microbiota, while fermentable fibre is metabolised by the gut bacteria to produce short-chain fatty acids that promote the proliferation of beneficial bacteria in the colon and also play a crucial role in risk reduction of NCDs [40,41]. Additionally, recent studies have highlighted the synergistic effect of phenolic compounds and rice bran dietary fibre on anti-hyperglycaemic activities [27,42,43].

Despite these health benefits, few studies have demonstrated the effect of dietary fibre on the texture and consumer acceptance of whole grain rice. The hardness of cooked whole grain rice containing dietary fibre is higher than that of cooked milled rice, which leads to a decrease in consumer acceptance of whole grain rice [4,12,17,44]. Additionally, Parween et al. [45] demonstrated that increased resistance starch content using genetic modification affects the textural property of rice, i.e., increased hardness and adhesiveness. In addition to dietary fibre, several factors also affect the rice texture among rice varieties, e.g., chemical composition, the amylose and protein content [46–48]; starch fine structure, such as the proportions of chain length and molecular size of amylose and amylopectin [49–51]; physicochemical properties, such as gelatinization temperature (GT) and viscosity [48,52–55], and physical properties, such as shape and size of the rice kernel [48,55]. Therefore, variation in the eating and cooking quality of rice is the most important aspect to be considered for improving rice variety to ensure customer satisfaction and health benefits. Surprisingly, a variation in the SDF to IDF ratio among varieties of whole grain rice was observed in this study, which is likely relevant to the textural properties of cooked whole grain rice. Our finding was consistent with the finding reported by Daou et al. [56], who demonstrated that SDF derived from defatted rice bran forms a viscous solution and increases viscosity, which positively correlates with the adhesiveness of cooked rice [54,57]. Furthermore, in a study by Mestres et al. [48], the amount of β-glucan, which is classified as SDF, was found to be negatively correlated with the hardness and chewiness of cooked milled rice.

Although a few reports explore the relationship between an increase in the dietary fibre content and the eating quality of rice, no report has been published on the influence of dietary fibre profiles on the textural properties of cooked whole grain rice. Therefore, this study is aimed at identifying the association between dietary fibre profiles and textural properties of cooked whole grain rice, which, in turn, can be utilised as a predictive indicator of the texture of whole grain rice and as a potential biomarker for breeding soft, whole grain rice to overcome consumer resistance.

Current methods for predicting both soluble and insoluble types of dietary fibre in whole grain rice accepted by the Association of Official Analytical Chemists (AOAC) International are expensive and time-consuming. The Prosky method, the enzymatic-gravimetric AOAC methods 985.29 and 991.43, provide easy determination of IDF and only high-molecular weight soluble dietary fibre (HMWSDF); however, the amount of SDF in whole grain rice is lower than the detection limit of this method [58]. Although the method developed by McCleary et al. [59], an enzymatic-gravimetry combined with high-performance liquid chromatography (HPLC) AOAC methods 2009.01 and 2011.25, can accurately determine the quantity of dietary fibre, including HMWSDF, low-molecular-weight soluble dietary fibre, and IDF; however, it is a complicated method. This enabled us to develop a simple predictive method for determining the dietary fibre content in whole grain rice by investigating the association of dietary fibre content in whole grain rice determined by the standard method and the alternative alkaline method. These findings provided vital information for improving the texture of high-nutrient rice.

2. Materials and Methods

2.1. Rice Varieties

Rice varieties (*Oryza sativa* L. ssp. *indica*) can be classified into six groups based on amylose content (AC) and pigment contents. In this study, ACs were subdivided into four groups: waxy (0–12%), low (12–20%), intermediate (20–25%), and high (25–33%), based on the classification provided by Juliano [60], Pang et al. [61], and the Waxy (*Wx*) gene provided by Liu et al. [62].

2.1.1. Non-Pigmented Rice

High-amylose rice: 66B09, Pinkaset4#20A09 (PK4#20A09), 16F35, MU2-00005, Pinkaset4#117A08 (PK4#117A08), and Pinkaset4#78A03 (PK4#78A03)

Intermediate-amylose rice: Doongara (DGR), M9997, Basmati (BMT), and Khaotahang (KTH)

Low-amylose rice: RD 43, RD 15, Pitsanulok 80 (PNL80), Pinkaset1 (PK1), Khaodawk-mali 105 (KDML105), Homsiam (HS), Hugdoi (HD), Pathumthani 1 (PTN1), Hommaliman (M7881), Sinlek (SL), and Homcholasid (HCS)

Waxy rice: Niewhomnuan (NHN) and RD 6

2.1.2. Pigmented Rice

Low-amylose rice: MU2-42, 909-10-3, Jaohomnil (JHN), Hom Lanna (HLN), Riceberry (RB), and Sungyodna (SYN)

Waxy rice: Klumhom (KH) and MU1-2313.

These rice varieties were provided by the Rice Science Center, Kasetsart University, Kamphaeng Saen Campus, Nakhon Pathom, Thailand. The whole grain rice samples were coarsely ground with a blender, followed by fine grinding and screening into particle sizes of 200 μm using a speed rotor mill, Pulverisette 14, Fritsch. The flour was stored in an airtight container at −20 °C until it was required for further analysis.

2.2. Rice Bran Fraction Quantification

The rice bran samples were separated by two methods, i.e., the milling method and the alkaline method. In the milling method, the bran was collected by milling whole grain rice in a rice polisher followed by roller milling. The percentage of bran removed from whole grain rice was expressed as the degree of milling (DOM), and the milled bran was calculated using the following equation provided by Gujral et al. [63] and Bautista and Siebenmorgen [64]:

$$Milled\ bran\ (\mathrm{g}/100\ \mathrm{g},\ dry\ basis) = \frac{Weight\ of\ (whole\ grain - milled)\ rice \times 100}{Weight\ of\ whole\ grain\ rice} \quad (1)$$

The alkaline method was adapted from the alkaline degradation test of rice endosperm, in which the starchy endosperm in rice kernel was digested by 1–7% (w/v) alkaline solution, depending on the alkaline-resistant properties [65,66]. To determine the weight of the bran layer without germ, rice germ was removed from whole grain rice by hand using cutting before incubation with an alkaline solution. In a brief process, 50 rice kernels with or without germ were immersed in 20 mL of potassium hydroxide (KOH) aqueous solution with different concentrations, depending on the alkaline-resistant properties as mentioned below, for 24 h at room temperature. Rice varieties with an alkaline spreading value (ASV) of more than 1 were treated with 3% (w/v) KOH solution, whereas rice varieties with ASV equal to 1 were treated with 6% (w/v) KOH solution. After a 24-h incubation, the rice endosperm starches were completely gelatinised and separated from the rice bran; then the detached bran layer with germ or without germ (Figure 1) was collected and washed with deionised water three times.

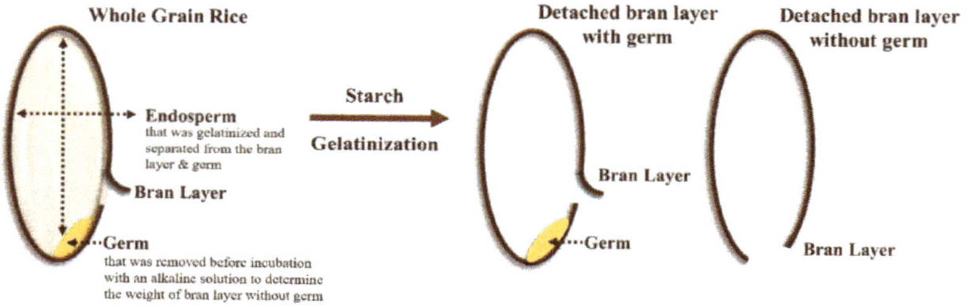

Figure 1. Separation of bran layer with germ or without germ from whole grain rice by the alkaline method.

To determine the dry weight of alkaline-treated rice bran samples, the samples were dried in an air oven at 105 °C for 24 h, and then weighed; the percentage of alkaline-treated bran layer with germ or without germ was calculated using the following equation:

$$BWG\ (\mathrm{mg/g}) = \left[\frac{Weight\ of\ bran\ layer\ with\ germ\ or\ without\ germ\ (\mathrm{g})}{Weight\ of\ whole\ grain\ rice\ (\mathrm{g})} \times 100\right] \times 10 \quad (2)$$

where BW is the percentage of bran layer weight with germ or without germ in whole grain rice (g/100 g, dry basis) and BWG is the bran layer weight with germ or without germ per gram of whole grain rice (mg/g, dry basis).

2.3. Dietary Fibre Analysis

Dietary fibre was determined by measuring carbohydrates that have a degree of polymerization (DP) of more than 2 and are not hydrolysed by the endogenous enzyme in the small intestine of humans. The enzymatic method based on AOAC methods 991.43 and 985.29 (K-TDFR, Megazyme) was used to estimate the dietary fibre content of whole grain rice and rice bran samples in this study. Before the analysis, defatted rice bran was produced using a modified version of the method given by Ren et al. [67]. Briefly, rice bran was extracted with cold acetone (1:10, w/v), followed by centrifugation. The supernatant was discarded, and the remaining pellets were re-extracted twice before air-drying under the hood overnight, followed by powdering. The sample was subjected to sequential enzymatic digestion by heat-stable α-amylase, protease, and amyloglucosidase. In a brief process, 1 g of whole grain rice flour or defatted bran powder was boiled for 30 min with 50 mL of 0.05 M MES/TRIS buffer (pH 8.2) and 0.2 mL of thermostable α-amylase (3000 Units/mL). After cooling, the solution was incubated at 60 °C with 0.1 mL of protease (50 mg/mL; ~350 tyrosine Units/mL). After 30 min of incubation, it was adjusted to pH 4.5 with 0.561 N HCl and further incubated at 60 °C for 16 h with 0.2 mL of amyloglucosidase (3300 Units/mL). Upon complete digestion, the solution was filtered to separate the insoluble (residue) and soluble (filtrate) fractions. IDF: The residue was washed with 78% ethanol, 95% ethanol, and acetone, then dried and weighed. The weight of the residue corrected for crude protein and ash formed the total quantity of IDF, which was calculated as the percentage of whole grain rice flour or rice bran powder. SDF: To deionise, the filtrate was further passed through a column packed with mixed-bed ion-exchange resin, following which the deionised solution was concentrated and filtered again through a 0.45 μm-membrane filter. The filtrate consisting of SDF was quantified by HPLC with a refractive index detector (Shimadzu RID-10A HPLC system, Shimadzu Corporation, Kyoto, Japan) based on Ohkuma's method [68] with modification and AOAC methods 2009.01 and 2011.25 (K-INTDF, Megazyme). The SDF was expressed as the percentage of whole grain rice flour or rice bran powder, whereas the total dietary fibre (TDF) was expressed as the sum of IDF and SDF.

2.4. Predictive Model for Dietary Fibre Content in Whole Grain Rice

A simple prediction model was developed in this study to ascertain the dietary fibre content in whole grain rice using linear regression between the bran layer fraction weight estimated by the alkaline method (Section 2.2) and the dietary fibre content determined by the standard method (Section 2.3). The actual values of dietary fibre content (Y): SDF, IDF, and TDF, were plotted against the bran layer weight without germ per whole grain (BW, X_1) (Figure S1A). The BW was calculated using the Equation (2). Additionally, to remove the difference in kernel size, the correlation of dietary fibre content with the bran layer weight based on the surface area of the rice kernel, representing bran thickness, was observed. The length (cm), width (cm), and thickness (cm) of whole grain rice were manually measured using a vernier calliper with 0.1 mm least count. The actual value of dietary fibre content (Y) was plotted against the bran layer weight without germ per surface area (BWS, X_2) (Figure S1B). The surface area and bran thickness of rice kernel were calculated using the equation provided by Nádvorníková et al. [69] and Chen and McClung [70], as shown below:

$$Srface\ area\left(s,\ \frac{\text{cm}^2}{\text{kernel}}\right) = \frac{\pi \times L^2 \times (W \times T)^{\frac{1}{2}}}{(2 \times L) - (W \times T)^{\frac{1}{2}}} \quad (3)$$

$$Surface\ area\left(S,\ \frac{\text{cm}^2}{\text{g}}\right) = \frac{s \times 50\ kernels}{FKW} \quad (4)$$

$$Bran\ thickness\left(BWS,\ \frac{\text{mg}}{\text{cm}^2}\right) = \frac{BWG}{S} \quad (5)$$

where s is the surface area of a single kernel (cm^2/kernel); S is the total surface area per gram of whole grain rice (cm^2/g); L, W, and T are the length, width, and thickness of kernel (cm), respectively; FKW is the dry weight of 50 kernels of whole grain rice (g); BWG is the bran layer weight per gram of whole grain rice (mg/g), and BWS is the bran layer weight per unit of surface area (mg/cm^2).

2.5. Chemical Composition Analysis

The AC was determined by the iodine-colourimetric method, based on the method used by Juliano [71]. The total fat was determined using Soxhlet extraction with petroleum ether based on AOAC method 945.16. The moisture content was measured by the gravimetric method based on the International Organization for Standardization (ISO) method 712:2009. Crude protein was determined by Kjeldahl analysis, according to the AOAC method 2001.11. Crude ash was determined by incineration at 525 °C for five hours according to AOAC method 942.05.

2.6. Alkaline Degradation Test

The alkaline degradation test was conducted by the method employed by Little et al. [65], with minor modifications to estimate GT and alkaline-resistant properties of whole grain rice samples. Eight kernels of whole grain rice were placed in a closed 10-cm Petri dish containing 20 mL of 1.7% (w/v) KOH aqueous solution for 24 h at room temperature. The spreading value was rated visually on a 7-point numerical scale (1 = intact; 7 = greatly dispersed), and the average scores of eight kernels were taken as the spreading value.

2.7. Texture Profile Analysis

Whole grain rice samples were soaked in water in a ratio of 1:2 in aluminium cups, then cooked in a stainless-steel streamer for 30–40 min until no white starch core could be observed before the analysis. A texture profile analysis (TPA) of cooked whole grain rice samples was conducted with a texture analyser (TA-XT plus, Stable Micro System, Godalming, UK), based on the method used by Parween et al. [45] and Guillen et al. [72], which demonstrated the significant correlation with sensory evaluation by trained panellists. A 50-mm cylinder probe was set at 30 mm above the base. The TPA force-deformation curve was obtained using a two-cycle compression with a force-versus-distance program. Three warm rice kernels were put onto the base platform under the centre of the probe and compressed to 90% of the original cooked grain thickness. Pre-test speed, test speed, and post-test speed were 1, 0.5, and 0.5 mm/sec, respectively. In total, nine measurements were performed for each sample (3 measurements per cup × 3 cups).

2.8. Radar Chart Image Creation

Two types of radar charts were developed to display the complex correlation among AC, SDF to IDF ratio, and textural characteristics of whole grain rice. The first one was designed to plot a series of the SDF to IDF ratio: low, medium, high, and very high over the average values of each TPA parameter (Figure 6A). The second one was developed to plot the values of TPA parameters—hardness, springiness, cohesiveness, and SDF to IDF ratio—across rice varieties with high-intermediated AC (Figure 6B) and with low AC (Figure 6C). The SDF to IDF ratio and TPA parameters were classified into four groups or 4-point scales using the following numerical rating: 1 = lowest and 4 = highest, as described in Table 1. Scaling was performed by dividing the difference between the maximum and minimum values in all samples by 4, based on the method described by Bernstein [73] with modification.

2.9. Statistical Analysis

The analysis was performed using Statgraphics Centurion XVII software (Statpoint Technologies, Warrenton, VA, USA). The data were analysed in triplicate by one-way analysis of variance, and Duncan's multiple range test was used to determine statistically significant differences among the samples. These differences were indicated by different letters in the columns when the p value was lower than 0.05. Linear regression was also analysed using Pearson correlation two-tailed test with a significance level of 0.05 and 0.01 [74]. A correlation matrix was developed to identify the correlations between two variables using a linear relationship Pearson correlation coefficient with R-statistic version 5.5.1 (The R Foundation for Statistical Computing, Vienna, Austria) at a significance level of 0.05 and 0.01 [75] and drawn with the DISPLAYR web free service.

Table 1. 4-Point rating scale of texture parameters and the ratio of SDF and IDF.

Parameters	Measurement	Definition		4-Point Scale			
				<1 Slightly	<2 Moderately	<3 Very	<4 Extremely
Hardness	Determined by the peak height of the first curve	The force required to compress the food sample	Hard (N)	0–10	10–20	20–30	30–40
Adhesiveness	Determined by negative force on the upstroke representing work to pull the plunger away from the sample	The degree to which the food sample sticks to the hand, mouth surface, or teeth	Sticky (N.s)	0–0.1	0.1–0.2	0.2–0.3	0.3–0.4
Springiness	Determined by the ratio of distance travelled by the plunger on the two curves	The degree to which the deformed food sample returns to its original size and shape relating to sample recovery after the first compression	Springy (s/s)	0–0.33	0.33–0.67	0.67–1.00	1.00–1.33
Cohesiveness	Determined by the ratio of the area under the second compression to the area under the first compression	The degree to which particles of food sample stick together	Cohesive (N.s/N.s)	0–0.25	0.25–0.50	0.50–0.75	0.75–1.00
Gumminess	Calculated by hardness × cohesiveness	The energy required to disintegrate the food sample until it is ready to be swallowed	Gummy (N)	0–5	5–10	10–15	15–20
Chewiness	Calculated by gumminess × springiness	The energy required to chew the food sample until it is ready to be swallowed	Chewy (N)	0–5	5–10	10–15	15–20
				4-Point scale/group			
SDF to IDF ratio				**<1** **Low**	**<2** **Med**	**<3** **High**	**<4** **Very high**
				0–0.16	0.16–0.28	0.28–0.40	0.40–0.54

Adapted from [73,76–79].

3. Results

3.1. Development of a Simple Prediction Method for Determining the Dietary Fibre Content in Whole Grain Rice Based on Bran Fraction Weight

In this study, the fraction weight of rice bran was determined using both the classical milling method in the preliminary part of this research and the alternative method modified from the alkali degradation test (Table S1). When the association between bran weight and dietary fibre was investigated, the results showed no correlation between the amount of TDF, IDF, or SDF in the whole grain rice and the rice bran weight determined by the milling method (Figure S2). Likely, the contamination of starchy endosperm in the milled rice bran caused the overestimation of bran weight [70]. To reduce the measurement error, an alternative method modified from the alkali degradation test was developed to determine the fraction weight of rice bran. Figure S1A shows that the percentage of bran layer without germ (BW), determined by the alkaline method, has a strong correlation with the percentage of IDF ($r = 0.81$, $p < 0.01$) and the percentage of TDF ($r = 0.75$, $p < 0.01$). A weak relationship ($r = 0.42$, $p < 0.05$) between SDF content in the whole grain and the bran layer was observed as expected. IDF was mostly localised in the bran layer, whereas SDF was distributed throughout the endosperm, which is described later.

Interestingly, Chen and McClung [70] uncovered the correlation between the physical traits and bran traits of whole grain rice. The authors of the study at hand further hypothesised that kernel size and surface area can influence the amount of dietary fibre in whole grain rice. Thus, the relationship between the percentage of dietary fibre and bran thickness (BWS, mg/cm^2) was also observed in this study. Bran thickness, expressed as bran weight independent of grain size (Tables S1 and S2), showed a strong correlation with the percentage of IDF and TDF ($r = 0.78$ and 0.69, $p < 0.01$, respectively), but no significant relationship was found with the percentage of SDF ($r = 0.32$), as presented in Figure S1B.

While both bran weight and kernel shape are related to the amount of dietary fibre in whole grain rice, dietary fibre content exhibiting a comparatively stronger relationship with BW than with BWS suggests that the former correlation is more accurate and practical for screening the dietary fibre content in a huge amount of whole grain rice samples. Thus, the percentage of bran layer can potentially be used to estimate the percentage of IDF and TDF in whole grain rice using the following linear regression model: % IDF = 0.73X + 1.09 and % TDF = 0.92X + 1.41, where X represents the BW determined by the alkaline method, whereas the percentage of SDF in whole grain rice was calculated by subtracting the percentage of TDF from IDF. Lastly, while the predicted values of IDF and TDF highly correlated with the value of IDF and TDF quantitated by the AOAC standard method, the predicted SDF value significantly correlated with the SDF value quantitated by the standard method (Figure 2).

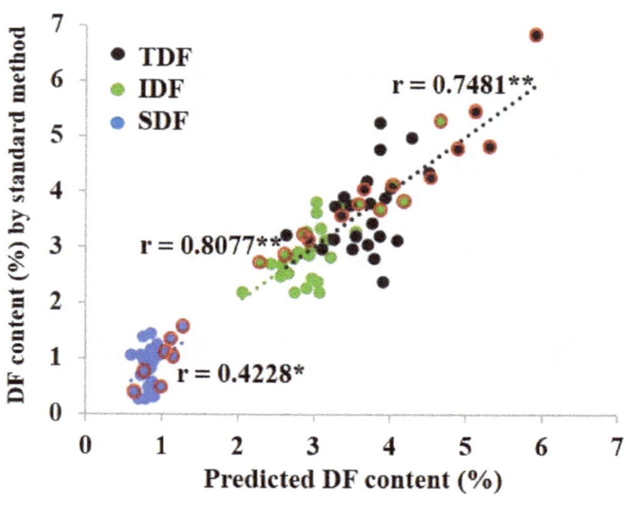

Figure 2. Linear regression between soluble dietary fibre (SDF), insoluble dietary fibre (IDF), and total dietary fibre (TDF); predicted value from the alkaline method and determined value from the standard method. Rice varieties with ASV equal to exactly 1 are shown as circles with a red border. * Correlation is significant at the 0.05 level. ** Correlation is significant at the 0.01 level.

3.2. Variation in the Distribution of Dietary Fibre in Whole Grain Rice

A slight variation in the dietary fibre composition was observed among a series of whole grain rice samples. Figure 3A indicates that the average value of SDF was 0.82%, varying from 0.27% to 1.44%; the average value of IDF was 2.97%, ranging from 2.18% to 3.82%, which is in line with that in the previous reports [13,17,36,80]. For non-waxy rice, the three highest values of both SDF and IDF were found in all low-AC and pigmented rice samples, SDF: HLN (1.24% SDF), RB (1.37% SDF), and MU2-42 (1.44% SDF); IDF: HLN (3.75% IDF), MU2-42 (3.81% IDF), and SYN (3.82% IDF) (Table S3). Additionally, the amount of both SDF and IDF in whole grain pigmented rice was observed to be significantly higher than that in non-pigmented rice (Figure 3C,D). The amount of dietary fibre in rice bran samples, separated from whole grain rice by milling, was also quantitated (Table S4). Figure 3B indicates that the average value of SDF was 2.62%, varying from 0.92% to 4.56%, whereas the average value of IDF was 30.81%, ranging from 25.25% to 39.71%, which is in line with that in published reports [13,56,81,82]. Although the TDF content in milled bran was considerably higher than that in whole grain, the major portion of dietary fibre in rice bran was insoluble, constituting about 90% of TDF. This suggests that only the insoluble type of dietary fibre is primarily concentrated in the outer layer of whole grain rice, while the soluble type is distributed throughout the endosperm of rice grain.

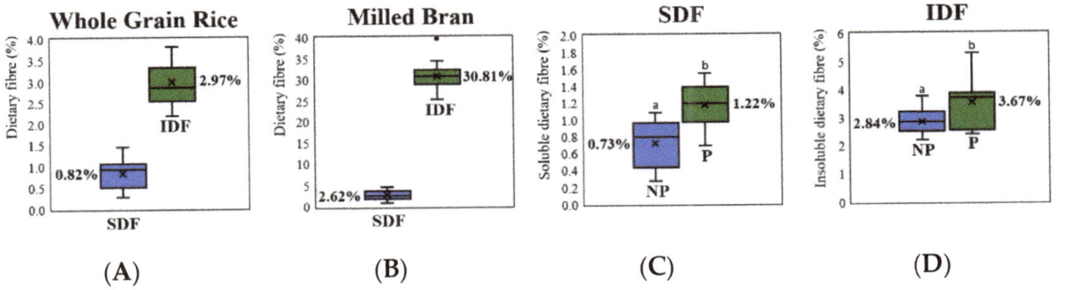

Figure 3. Overview of variation in SDF and IDF distribution (**A**) in whole grain rice samples varying from high amylose content (AC) to low AC, expressed as the percentage of total grain weight and (**B**) in milled bran samples expressed as the percentage of rice bran powder; the difference in (**C**) SDF or (**D**) IDF of whole grain non-pigmented (NP) and pigmented (P) rice, expressed as the percentage of total grain weight. Values with different letters are significantly different with $p < 0.05$.

When the SDF to IDF ratio of whole grain rice was compared among the rice varieties, a wide variation was found, ranging from 0.1 (PK+4#20A09, 26.86% AC; PK+4#117A08, 26.69% AC; KTH, 21.73% AC, and RD43, 19.98% AC) to 0.5 (SL, 14.86% AC and RB, 13.96% AC) (Figure 4A, Table S3). Conversely, a slight difference was found in the SDF to IDF ratio of milled bran samples, ranging from 0.03 to 0.18 (Figure 4B, Table S4). This suggests a variation in SDF distribution throughout the endosperm among rice varieties and that rice with lower AC exhibits a higher SDF to IDF ratio. Thus, the distribution of SDF in rice endosperm likely also influences the hardness of cooked whole grain rice.

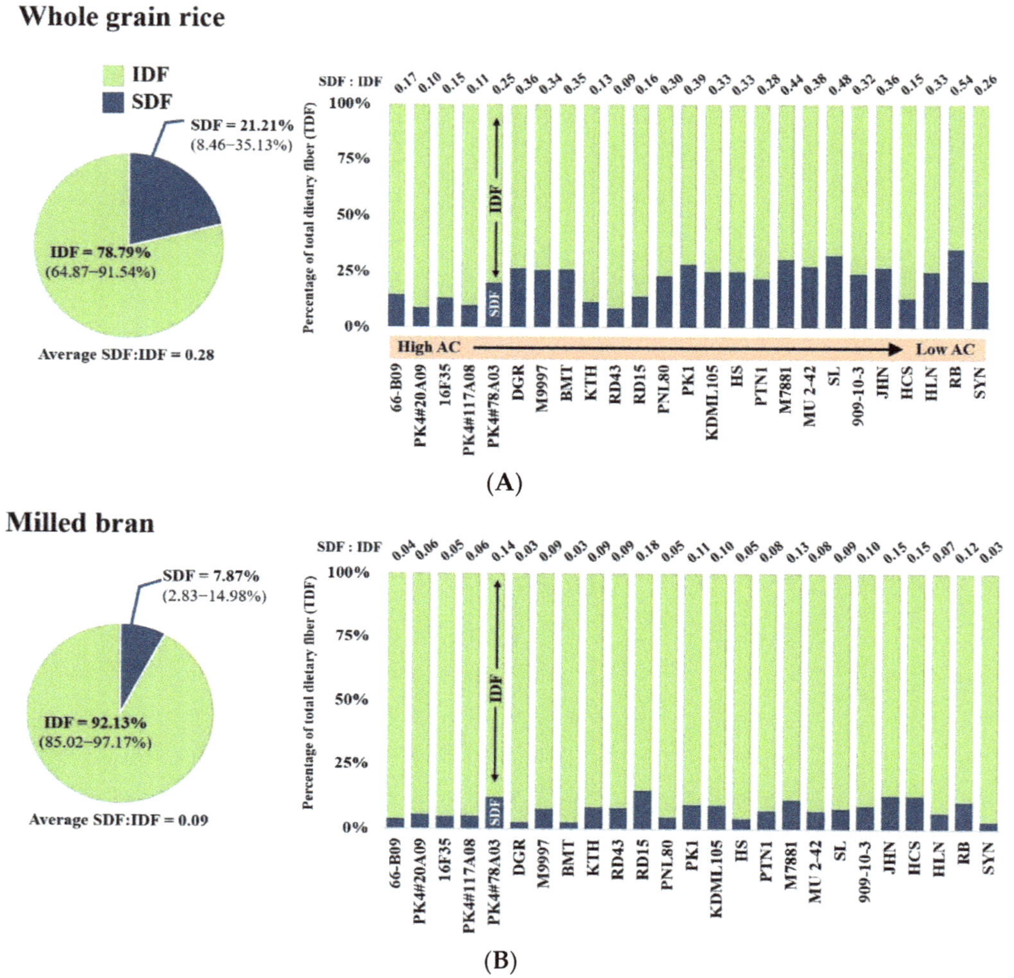

Figure 4. Variation in dietary fibre profiles, SDF and IDF of (**A**) whole grain rice samples with different ACs and (**B**) milled bran samples, expressed as a percentage of TDF.

3.3. *Correlations of Dietary Fibre Profiles, Textural Characteristics, and Amylose Content of Whole Grain Rice*

Several lines of evidence revealed the influence of factors such as chemical composition [46,47], starch fine structure [49,50], and physicochemical properties [52–55] on the textural properties of whole grain rice. Moreover, an association between the SDF to IDF ratio and the textural characteristics of whole grain rice with differing AC was discovered in this study. This led us to analysing the associations among different properties of whole grain rice, such as dietary fibre profiles, AC, GT, and textural characteristics using multiple regression analysis. The AC showed a strong positive correlation with most of the TPA parameters but a negative correlation with adhesiveness (Figure 5, Table S5). For the dietary fibre profile, variation in dietary fibre had a strong influence on TPA and the eating quality of whole grain rice. In particular, SDF and SDF to IDF ratio of whole grain rice contributed strongly to the softness of cooked whole grain and to most of the TPA parameters (Figure 5). Therefore, IDF in whole grain rice gave no significant contribution to cooked whole grain rice in the selected rice germplasm. Moreover, the hardness and gumminess of cooked whole grain rice were strongly negatively correlated with the SDF content ($r = -0.70$ and -0.72, respectively) and the SDF to IDF ratio ($r = -0.74$ and -0.69, respectively) in cooked whole grain rice at a 99% confident level; cohesiveness and chewiness were moderately negatively correlated with the SDF content ($r = -0.58$ and -0.62, $p < 0.01$, respectively) and the SDF to IDF ratio ($r = -0.45$, $p < 0.05$ and $r = -0.55$, $p < 0.01$, respectively). Conversely, the SDF content in whole grain rice was moderately positively correlated with the adhesiveness of cooked whole grain rice ($r = 0.45$, $p < 0.05$). All these findings indicated that the distribution of SDF throughout the rice endosperm highly reduced the hardness of cooked rice, whereas AC increased its hardness.

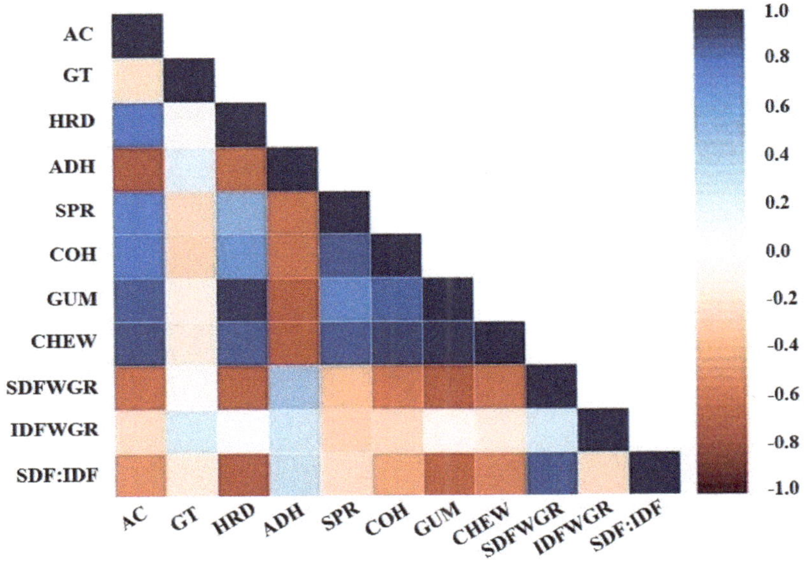

Figure 5. Pearson's correlation matrix: correlation between amylose content, gelatinization temperature, textural parameters, and dietary fibre profiles. AC = amylose content; GT = gelatinisation temperature; HRD = hardness; ADH = adhesiveness; SPR = springiness; COH = cohesiveness; GUM = gumminess; CHEW = chewiness; SDFWGR = SDF in whole grain rice; IDFWGR = IDF in whole grain rice; SDF:IDF = SDF to IDF ratio of whole grain rice.

3.4. *Influence of Dietary Fibre Profiles on the Softness of Whole Grain Rice*

To further observe the association between the textural properties and the dietary fibre profiles of whole grain rice, cooked whole grain rice samples were subjected to TPA conducted using a texture analyser with a two-cycle compression, which mimics the first and second bites on a rice sample, for predicting the texture of whole grain rice. The texture characteristics of all cooked whole grain rice samples have been described in Table S6. The RB rice with the highest SDF to IDF ratio showed the lowest value of hardness and gumminess (13.04 N and 4.47 N, respectively), whereas PK+4#20A09 rice with the lowest SDF to IDF ratio demonstrated the highest value of the above textural parameters (36.67 N and 16.00 N, respectively). Conversely, RB or PK+4#20A09 rice did not show the highest or lowest value for other textural parameters, i.e., chewiness (2.09–11.52 N), adhesiveness (10.62–77.14 mN.s), springiness (0.46–0.89 s/s), and cohesiveness (0.31–0.53 N.s/N.s). Only hardness and gumminess showed a strong correlation with the SDF to IDF ratio of whole grain rice.

Furthermore, whole grain rice samples were grouped by the SDF to IDF ratio into a low ratio (<0.16), medium ratio (0.16–0.28), high ratio (0.28–0.40), and very high ratio

(0.40–0.54), to associate with expected textural characteristics (Figure 6A). Based on the textural parameters (Table 1), the association between dietary fibre and the texture of cooked whole grain rice was established. Rice with a lower SDF to IDF ratio was harder, gummier, and chewier than rice with a higher SDF to IDF ratio. However, no statistical differences were detected in adhesiveness ($p = 0.31$) and springiness ($p = 0.45$) among the groups of cooked whole grain rice with different SDF to IDF ratios (Figure 6A). Thus, the SDF to IDF ratio of whole grain rice negatively correlates with the hardness, gumminess, chewiness, and cohesiveness of cooked whole grain rice.

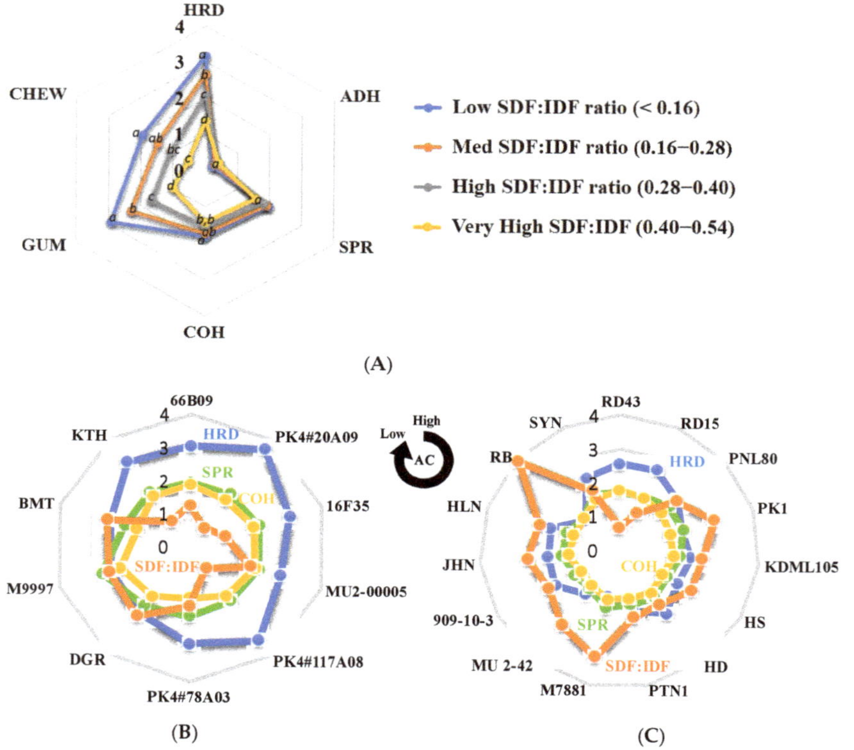

Figure 6. Radar charts showing (**A**) textural characteristics of whole grain rice containing different SDF:IDF ratios and (**B**,**C**) relationships between textural properties and dietary fibre profiles of (**B**) whole grain Wx^a rice with AC higher than 20% and (**C**) whole grain Wx^b rice with AC lower than 20%. Rice varieties are sorted in a clockwise direction from high to low AC. The texture parameters including hardness (HRD), adhesiveness (ADH), springiness (SPR), cohesiveness (COH), gumminess (GUM) and chewiness (CHEW), and SDF to IDF ratio were expressed as a 4-point scale. Values with different letters are significantly different with $p < 0.05$.

Nevertheless, the *Wx* gene plays important roles in regulating grain AC and cooked rice quality [83]. Two key polymorphisms GT and TT, identified at the 5′ splice site of the first intron in the 5′ untranslated region, define two predominant *Wx* alleles, namely Wx^a and Wx^b. The Wx^a rice contains the GT haplotype, exhibiting intermediate to high AC, whereas the Wx^b rice contains the TT haplotype, exhibiting low AC [62]. In this study, all rice samples were genotyped using the GT/TT single-nucleotide polymorphisms to determine *Wx* gene haplotypes in addition to amylose analysis (Table S7). To understand the relationship among AC, SDF to IDF ratio, and textural properties of whole grain rice, we created a radar chart analysis to illustrate such complex relationships across diverged rice varieties with high, intermediate, and low AC (Figure 6B,C).

Among the high-intermediate AC group (Wx^a group), rice with lower SDF to IDF ratios had a highly significant correlation with hardness ($r = -0.95$, $p < 0.01$). Interestingly, rice varieties in the intermediate AC group (20–25% AC) had a higher average SDF to IDF ratio than those in the high AC rice group. As expected, cooked whole grain rice from the high AC rice group was harder than that from the intermediate AC rice group due to both AC and SDF to IDF ratio. Within the intermediate AC group, SDF to IDF ratio played a major role in the softness of cooked whole grain rice. Whole grain rice from DGR (25.01% AC), M9997 (23.91% AC), and BMT (22.48% AC) with high SDF to IDF ratio cooked softer than that from KTH (21.73% AC) (Figure 6B). Particularly, among the low AC rice group, SDF to IDF ratio played a sensitive role in determining the softness of cooked whole grain rice (Figure 6C). There were a wide range of ACs and SDF to IDF ratios among the selected varieties (Table S3). However, SDF to IDF ratio had a strongly negative correlation with hardness ($r = -0.91$, $p < 0.01$). The most contrasting varieties on dietary fibre profile, hardness, and AC are RB and RD43 (Tables S3 and S6). Recorded as the richest SDF, RB had also the highest SDF to IDF ratio and the softest cooked whole grain rice. Conversely, RD43 had the highest AC and hardness containing the lowest SDF content and SDF to IDF ratio among the low AC rice group. These results suggest that reductions in the SDF to IDF ratio of whole grain rice increase the hardness of cooked rice. Thus, in addition to impacting AC, the SDF to IDF ratio also influences the textural characteristics of cooked whole grain rice.

4. Discussion

4.1. More Accuracy in the Alternative Alkaline Method for Estimation of Dietary Fibre

Current methods for measuring both soluble and insoluble types of dietary fibre in whole grain rice—the enzymatic-gravimetric method combined with the HPLC method based on AOAC methods 2009.01 and 2011.25—are expensive and complicated. Here, we found that the amount of dietary fibre in milled rice is approximately half of that in whole grain rice (Table S8), consistent with a previous study that demonstrated that the TDF values of milled rice with low and high AC were 59% and 49% of the values found in whole grain rice, respectively [11]. This suggests that the other half of dietary fibre is located in the bran fraction. Dietary fibre is the second largest component of rice bran [29,84,85]. Consequently, we had to investigate the association between the fraction weight of rice bran and the amount of dietary fibre in whole grain rice for developing the potential model to predict the dietary fibre content in whole grain rice. To date, bran weight has been estimated from the weight lost during the milling process of whole grain rice [63,64]. However, we have observed that rice bran fractions prepared by the milling method are contaminated, with variable amounts of starch ranging from 6.8% to 35.1% due to kernel size and thickness [63], DOM [64], and type of milling processes used [86,87]. To reduce overestimation, the total starch concentration in the bran was determined and subtracted from the milled bran weight [70,88]. Due to a measurement error in the milling method, we developed the modified alkali disintegration method to provide a more accurate value of rice bran that was further used for predicting the dietary fibre.

The alkali degradation test, also referred to as the alkaline spreading method, was employed as an indirect quality assessment of GT [65,89,90]. During alkali spreading, KOH gelatinises starch (particularly, its amorphous region), causing degradation of the long, linear, and branched chains of amylose and amylopectin, and resulting in rice grain gelatinisation [90]. Rice bran, which is mainly composed of fibre, lipids, and protein, can be separated from the starchy endosperm fraction. However, some lipids, proteins, and arabinoxylan (AX) are partly dissolved in the alkaline solution during separation [91,92]. The alkaline solvent also derives solubilised AX from the cell wall matrix by the disruption of hydrogen and covalent bonds, thereby resulting in the loss of alkali-solubilised AX during the washing of rice bran [93,94]. This explains the reason for the lower percentage of bran layer determined by the alkaline method compared to the percentage previously published [16,85].

We demonstrated that the predicted values of IDF and TDF in whole grain indica rice, calculated from bran layer fraction weight determined by the alkaline method, were highly correlated with those found by the AOAC standard method, whereas the predicted SDF value in whole grain indica rice was weakly correlated with the analysed value by the standard method due to distribution of SDF throughout the whole grain. However, the alkaline method can provide a more accurate value of SDF than the milling method; moreover, it is cheaper and more simplified compared to the enzymatic-HPLC standard AOAC method. The gravimetric AOAC method does not provide an accurate SDF value due to a small amount of SDF in whole grain rice. Furthermore, chain length distribution of amylopectin is known to affect GT [95]. In this study, rice varieties with high GT were included and we found that the relationship between the predicted value from the alkaline method and the determined value from the standard method of rice varieties with an ASV equal to 1 showed the same result as that of rice varieties having an ASV of more than 1 (Figure 2, Table S2). This suggests that treating rice samples with different concentrations of KOH (3–6%) does not disturb the estimation of dietary fibre content in whole grain rice.

Additionally, the correlation between rice bran composed of bran layer with germ and dietary fibre of whole grain rice was considered. The result showed that the relationship between rice bran weight (BW, g/100 g) or bran thickness (BWS, mg/cm^2) (Tables S1 and S2) and the percentage of SDF ($r = 0.36$, $r = 0.24$, respectively) or IDF ($r = 0.73$, $r = 0.67$, $p < 0.01$, respectively) or TDF ($r = 0.62$, $r = 0.56$, $p < 0.01$, respectively) was a bit weaker than that of the bran layer without germ. A possible explanation for this result is the difference in chemical compositions between the bran layer and germ; the germ is composed of a lower dietary fibre than the bran layer [24]. As all samples were selected mainly from elite indica rice varieties, the linear relationships between BW or BWS and SDF are more predictive for long-slender-grain indica rice varieties than short-rounded-grain japonica varieties. Interestingly, the BW and BWS of whole grain pigmented rice were higher than those of the non-pigmented rice [70]. This is consistent with our findings, showing that the amount of IDF, which has a strong correlation with either BW or BWS, was significantly higher in whole grain pigmented rice than in non-pigmented rice.

4.2. Distribution of Soluble Dietary Fibre throughout Rice Endosperm

The comparison of the SDF to IDF ratio in whole grain and rice bran in this study indicates that the majority of SDF and about half of IDF are distributed throughout the endosperm of the rice grain, while the remaining small amount of SDF and the other half of IDF are concentrated in the bran layer of whole grain rice. This finding is consistent with that in a published report, which showed that the values of SDF, IDF, and TDF in milled rice are 67%, 49%, and 53%, respectively, of those values in whole grain rice [17]. Further, a variation was found between SDF distribution among rice varieties, and the low AC rice had a higher SDF to IDF ratio. Most dietary fibre in cereal grains is derived from the cell wall material [96,97], and recent studies, using monosaccharide analysis, have proposed that the composition of cell wall-derived dietary fibre in milled rice comprises glucan, pectin, arabinogalactan, and glucurono (arabino)xylan [98,99]. Meanwhile, other studies have reported that the profile of non-starch polysaccharides in whole grain rice and milled rice is composed of cellulose, AX, pectin, fructan, β-glucan, and resistant starch [100,101].

Only limited information is available on the composition of SDF in whole grain rice with different ACs. Therefore, we also investigated the composition of SDF including soluble AX, β-glucan, and pectin in whole grain rice with different ACs and rice bran samples (Table S9). The AX is among the major hemicellulosic components in cereal grain cell walls, and its structure comprises a linear backbone of β-(1–4)-linked xylose residues with arabinose substitution at the O-2 and O-3 positions. β-glucan is a water-soluble dietary fibre composed of glucose monomers linked together via β-(1–4) and β-(1–3) glycosidic bonds. Pectin is the most complex polysaccharide in plant cell walls, composed of nearly

70% galacturonic acid covalently linked at the O-1 and O-4 positions [31,102,103]. The results showed that the average percentages of soluble AX, β-glucan, and pectin were 7%, 11%, and 28%, respectively, of the SDF in whole grain rice, accounting for about 46% of the SDF content in whole grain rice. According to other published reports [98–101], the other half of SDF in whole grain rice might be resistant starch, arabinogalactan, and fructan. Interestingly, a greater variation was observed in β-glucan and pectin content in whole grain rice compared to that in rice bran; the highest amount of β-glucan and pectin was found in low AC rice, while the lowest amount was observed in high AC rice. This suggests that the distribution of β-glucan and pectin throughout the endosperm in low AC rice is higher than that in high AC rice, which is consistent with a previous report [48].

4.3. SDF to IDF Ratio as a Potential Biomarker for Selecting Eating Quality of Whole Grain Rice

Higher intake of whole grain rice is associated with a lower risk of NCDs [13,18,21–25], the reason being the high concentration of bioactive compounds in bran and germ fraction, e.g., phytochemicals and dietary fibre. They play various roles in biological activities, such as anti-oxidant, anti-diabetic, anti-obesity and cholesterol-lowering, anti-cancer, and anti-inflammatory activities [15,22–29]. Consumer preference regarding eating and cooking qualities is a strategic goal to achieve consumer acceptance in rice breeding. Eating and cooking quality, including water uptake, cooking temperature, grain size and shape, aroma, and texture, is mainly controlled by physicochemical properties, such as gelatinisation, retrogradation and pasting properties, the molecular structure of starch, and other nutritional compositions in rice kernel [104,105]. Mir et al. [1] revealed that consumers globally tend to prefer soft-textured white rice, which highly correlates with a high GI [10–12,34,35] and a high risk of developing T2D [6–9]. Recently published reports have demonstrated that an increase in the hardness of rice is associated with lower consumer acceptability [2–4]. The hardness parameter constitutes the force required to bite through the rice with molars, and chewiness implies the amount of work required to chew the rice until it is ready to swallow, which also predicts the hardness of the rice. Meanwhile, adhesiveness is interpreted as the mouthfeel of stickiness, i.e., the degree to which the food sample sticks to the hand, mouth surface, or teeth; cohesiveness indicates the degree to which the rice deforms rather than cracks when bitten by molars [76]. Here, the textural properties of cooked rice samples were determined by an instrumental texture analyser; however, previous reports [45,72] have demonstrated a significant relationship of rice texture attributes such as hardness, cohesiveness, and adhesiveness, between sensory evaluation by trained panellists and instrumental texture analyser under the same conditions used in this study. This suggests that the instrumental texture analyser has the potential to assist rice breeders to select the preferred cooked rice texture, in this case, the whole grain rice quality.

Carbohydrate structure, especially ACs, has a strong influence on the textural properties of whole grain rice [46,47]. Particularly, the proportions of chain length, DP, GT, and molecular size of amylose and amylopectin contribute to the hardness and stickiness of cooked milled rice [49–51,55]. An in-depth study on cooked rice quality of whole grain rice has been overlooked due to low marketing demand. We consider whole grain rice as a practical solution for rice biofortification. Despite the many nutritional benefits of whole grain rice, its low palatability induces resistance in consumers. Here, we dissect the key roles of dietary fibre profile in eating quality of whole grain rice. There have been reports on the impact of TDF on the hardness of cooked whole grain rice [4,12,17,44]. In this study, we demonstrated that an increase in the amount of SDF in whole grain rice decreases the hardness, cohesiveness, gumminess, and chewiness but increases the adhesiveness of cooked whole grain rice. Rice varieties in the intermediate AC group have a higher average SDF to IDF ratio than the high AC rice group. Interestingly, among the intermediate AC groups or low AC groups, cooked rice with a higher SDF to IDF ratio had a softer texture than cooked rice with a lower SDF to IDF ratio. This shows that the SDF to IDF ratio can determine the hardness among the intermediate AC rice group and low AC rice group. A possible reason for increased SDF content to reduce the hardness of cooked whole grain rice is the viscous properties of SDF. It is well known that viscosity or gel formation is one of the significant properties associated with SDF [33,106]. Previous studies have demonstrated that SDF can form a viscous solution and increase the solution viscosity [56], and Chen et al. [54] demonstrated a significant positive correlation between the adhesiveness and the viscosity of cooked rice. However, the impact of β-glucan on the hardness and chewiness of cooked milled rice was moderate [48]. β-glucan plays a crucial role in fighting against CVD, dyslipidaemia, insulin resistance, and obesity due to its fermentability and viscous properties [107]. Moreover, β-glucan can enhance the immune system via interactions with immune cells [108]. Some studies have also shown the antioxidant and prebiotic properties of β-glucan extracts of rice bran [109,110]. Despite the lack of information regarding the health benefits of pectin in rice, several reports have revealed that pectin has multiple positive effects on human health by the reduction of post-prandial glycaemic responses and the maintenance of normal blood cholesterol concentration, owing to its viscosity [96]. In this study, β-glucan and pectin constituted only a small fraction of SDF in selected varieties of whole grain rice. However, the benefits of whole grain rice are well documented in lowering the risk of NCDs and enhancing the immune system via phytoceutical compounds such as polyphenol, antioxidants, anthocyanin, and proanthocyanin [15,18–23]. We determined further that not only the SDF content but also the IDF content played crucial roles in the TPA of cooked whole grain rice. We have shown that the SDF to IDF ratio has a stronger link than SDF alone for precision breeding for the palatability of whole grain rice among varieties of cultivated tropical indica rice.

5. Conclusions

This study investigated the effects of dietary fibre profiles on the textural properties of cooked whole grain rice. Despite a slight variation in the dietary fibre composition of whole grain rice, the variation of SDF to IDF ratio in whole grain rice impacted the texture of cooked rice. Furthermore, this study demonstrated that the SDF to IDF ratio of whole grain rice was negatively correlated with hardness, cohesiveness, gumminess, and chewiness ($p < 0.01$) but positively correlated with the adhesiveness ($p < 0.05$) of cooked whole grain rice. This finding is helpful for future trends to improve softness and consumer acceptance of whole grain rice in indica rice (Figure 7). Furthermore, this study successfully developed a simplified approach to precisely predict dietary fibre profiles into fractions of whole grain rice using an alternative alkaline method that is practical for high-throughput screening of dietary fibre in precision rice breeding.

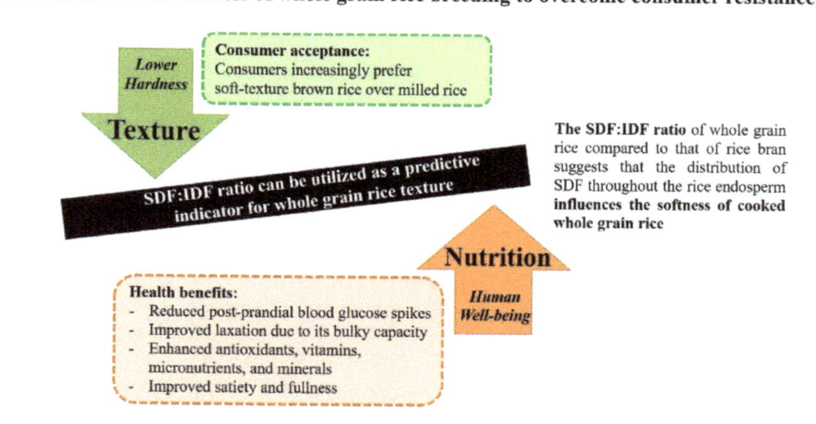

Figure 7. Conceptual diagram summarising the framework of this research.

Supplementary Materials: The supplementary material for this article can be found online at https://www.mdpi.com/article/10.3390/foods12040899/s1, Figure S1: Linear regression between the dietary fibre content and the bran layer weight determined by the alkaline method; Figure S2: Linear regression between the dietary fibre content and the bran weight determined by the milling method; Table S1: Rice bran traits of whole grain rice samples; Table S2: Physical characteristics of whole grain rice samples; Table S3: Chemical compositions of whole grain rice samples; Table S4: Chemical compositions of milled bran samples; Table S5: Pearson's correlation matrix; Table S6: Texture profile analysis of cooked whole grain rice samples; Table S7: Genotype of rice samples; Table S8: Range of basic nutrient compositions of whole grain rice, milled rice, and rice bran; Table S9: Comparison of the β-glucan and pectin content in whole grain rice and rice bran.

Author Contributions: S.W. and A.V. conceived and designed the work. A.V. and W.K. supervised the research. S.W., R.W. and E.C. performed the experiments and conducted the statistics data analysis. S.W. and A.V. wrote and revised the manuscript. All authors contributed to the article. All authors have read and agreed to the published version of the manuscript.

Funding: This work was supported by National Science Research and Innovation Fund (NSRF) via the Program Management Unit for Human Resources and Institution Development, Research and Innovation (Grant No. B16F630088); Distinguished Research Professor Grant, The National Research Council of Thailand; National Science and Technology Development Agency (NSTDA) (Grant No. P-18-52711). SW was financially supported by Postdoctoral Fellowship Program from Kasetsart University Research and Development Institute.

Data Availability Statement: The original contributions presented in this study are included in the article and Supplementary Materials. Further inquiries can be directed to the corresponding authors.

Conflicts of Interest: The authors declare no conflict of interest.

References

1. Mir, S.A.; Shah, M.A.; Bosco, J.D.; Sunooj, K.V.; Farooq, S. A review on nutritional properties, health aspects, shelf life and consumption of brown rice in comparison to white rice. *Cereal Chem.* **2020**, *97*, 895–903. [CrossRef]
2. Suwansri, S.; Meullenet, J.F.; Hankins, J.; Griffin, K. Preference maping of domestic /imported jasmine rice for US-Asian consumers. *J. Food Sci.* **2002**, *67*, 2420–2431. [CrossRef]
3. Maleki, C.; Oliver, P.; Lewin, S.; Liem, G.; Keast, R. Preference mapping of different water-to-rice ratios in cooked aromatic white jasmine rice. *J. Food Sci.* **2020**, *85*, 1576–1585. [CrossRef] [PubMed]
4. Gondal, T.A.; Keast, R.S.J.; Shellie, R.A.; Jadhav, S.R.; Gamlath, S.; Mohebbi, M.; Liem, D.G. Consumer acceptance of brown and white rice varieties. *Foods* **2021**, *10*, 1950. [CrossRef] [PubMed]
5. Se, C.H.; Khor, B.H.; Karupaiah, T. Prospects in development of quality rice for human nutrition. *Malays. Appl. Biol.* **2015**, *44*, 1–31.
6. Villegas, R.; Liu, S.; Gao, Y.T.; Yang, G.; Li, H.; Zheng, W.; Shu, X.O. Prospective study of dietary carbohydrates, glycemic index, glycemic loads, and incidence of type 2 diabetes mellitus in middle-aged Chinese women. *Arch. Intern. Med.* **2007**, *167*, 2310–2316. [CrossRef]
7. Nanri, A.; Mizoue, T.; Noda, M.; Takahashi, Y.; Kato, M.; Inoue, M.; Tsugane, S. Rice intake and type 2 diabetes in Japanese men and women: The Japan public health center-based prospective Study. *Am. J. Clin. Nutr.* **2010**, *92*, 1468–1477. [CrossRef]
8. Sun, Q.; Spigelman, D.; van Dam, R.B.; Holmes, M.D.; Malik, V.S.; Willett, W.C.; Hu, F.B. White rice, brown rice, and risk of type 2 diabetes in US men and women. *Arch. Intern. Med.* **2010**, *170*, 961–969. [CrossRef]
9. Hu, E.A.; Pan, A.; Malik, V.; Sun, Q. White rice consumption and risk of type 2 diabetes: Meta-analysis and systematic review. *Br. Med. J.* **2012**, *344*, e1454. [CrossRef]
10. Trinidad, T.P.; Mallillin, A.C.; Encabo, R.R.; Sagum, R.S.; Felix, A.D.R.; Juliano, B.O. The effect of apparent amylose content and dietary fibre on the glycemic response of different varieties of cooked milled rice and brown rice. *Int. J. Food Sci. Nutr.* **2013**, *64*, 89–93. [CrossRef]
11. Pletsch, E.A.; Hamaker, B.R. Brown rice compared to white rice slows gastric emptying in humans. *Eur. J. Clin. Nutr.* **2017**, *72*, 367–373. [CrossRef]
12. Huang, M.; Li, X.; Hu, L.; Xiao, Z.; Chen, J.; Cao, F. Comparing texture and digestion properties between white and brown rice of indica cultivars preferred by Chinese consumers. *Sci. Rep.* **2021**, *11*, 19054. [CrossRef] [PubMed]
13. Juliano, B.O. Rice: Overview. In *Encyclopedia of Food Grains*, 1st ed.; Wrigley, C., Corke, H., Seetharaman, K., Faubion, J., Eds.; Academic Press: Oxford, UK, 2016; pp. 125–129. [CrossRef]
14. Reddy, C.K.; Kimi, L.; Haripriya, S.; Kang, N. Effects of polishing on proximate composition, physicochemical characteristics, mineral composition and antioxidant properties of pigmented rice. *Rice Sci.* **2017**, *24*, 241–252. [CrossRef]

15. Saleh, A.S.M.; Wang, P.; Wang, N.; Yang, L.; Xiao, Z. Brown rice versus white rice: Nutritional quality, potential health benefits, development of food products, and preservation technologies. *Compr. Rev. Food Sci. Food Saf.* **2019**, *18*, 1070–1096. [CrossRef] [PubMed]
16. Zhao, M.; Lin, Y.; Chen, H. Improving nutritional quality of rice for human health. *Theor. Appl. Genet.* **2020**, *133*, 1397–1413. [CrossRef]
17. Carcea, M. Value of wholegrain rice in a healthy human nutrition. *Agriculture* **2021**, *11*, 720. [CrossRef]
18. Mohan, V.; Anjana, R.M.; Gayathri, R.; Bai, M.R.; Lakshmipriya, N.; Ruchi, V.; Balasubramaniyam, K.K.; Jakir, M.M.; Shobana, S.; Unnikrishnan, R.; et al. Glycemic index of a novel high-fiber white rice variety developed in India—A randomized control trial study. *Diabetes Technol. Ther.* **2016**, *18*, 164–170. [CrossRef]
19. Panlasigui, L.N.; Thompson, L.U. Blood glucose lowering effects of brown rice in normal and diabetic subjects. *Int. J. Food Sci. Nutr.* **2006**, *57*, 151–158. [CrossRef]
20. de Munter, J.S.L.; Hu, F.B.; Spiegelman, D.; Franz, M.; van Dam, R.M. Whole grain, bran, and germ intake and risk of type 2 diabetes: A prospective cohort study and systematic review. *PLoS Med.* **2007**, *4*, e261. [CrossRef]
21. Ye, E.Q.; Chacko, S.A.; Chou, E.L.; Kugizaki, M.; Liu, S. Greater whole grain intake is associated with lower risk of type 2 diabetes, cardiovascular disease, and weight gain. *J. Nutr.* **2011**, *142*, 1304–1313. [CrossRef]
22. Nakamura, S.; Ikeuchi, T.; Araki, A.; Kasuga, K.; Watanabe, K.; Hirayama, M.; Ito, M.; Ohtsubo, K. Possibility for Prevention of Type 2 Diabetes Mellitus and Dementia Using Three Kinds of Brown Rice Blends after High-Pressure Treatment. *Foods* **2022**, *11*, 818. [CrossRef] [PubMed]
23. Watanabe, S.; Hirakawa, A.; Nishijima, C.; Ohtsubo, K.; Nakamura, K.; Beppu, S.; Tungtrakul, P.; Quin, S.J.; Tee, E.; Tsuno, T.; et al. The new concept of "medical rice". *Adv. Food Technol. Nutr. Sci. Open J.* **2016**, *2*, 38–50. [CrossRef]
24. Moongngarm, A.; Daomukda, N.; Khumpika, S. Chemical compositions, phytochemicals, and antioxidant capacity of rice bran, rice bran layer, and rice germ. *APCBEE Procedia* **2012**, *2*, 73–79. [CrossRef]
25. Forster, G.M.; Raina, K.; Kumar, A.; Kumar, S.; Agarwal, R.; Chen, M.H.; Bauer, J.E.; McClung, A.M.; Ryan, E.P. Rice varietal differences in bioactive bran components for inhibition of colorectal cancer cell growth. *Food Chem.* **2013**, *141*, 1545–1552. [CrossRef]
26. Goufo, P.; Trindade, H. Rice antioxidants: Phenolic acids, flavonoids, anthocyanins, proanthocyanidins, tocopherols, tocotrienols, γ-oryzanol, and phytic acid. *Food Sci. Nutr.* **2014**, *2*, 75–104. [CrossRef] [PubMed]
27. Zhang, X.; Dong, L.; Jia, X.; Liu, L.; Chi, J.; Huang, F.; Ma, Q.; Zhang, M.; Zhang, R. Bound phenolics ensure the anti-hyperglycemic effect of rice bran dietary fiber in db/db mice via activating the insulin signaling pathway in skeletal muscle and altering gut microbiota. *J. Agric. Food Chem.* **2020**, *68*, 4387–4398. [CrossRef]
28. Mbanjo, E.G.M.; Kretzschmar, T.; Jones, H.; Ereful, N.; Blanchard, C.; Boyd, L.A.; Sreenivasulu, N. The genetic basis and nutritional benefits of pigmented rice grain. *Front. Genet.* **2020**, *11*, 229. [CrossRef]
29. Pereira, C.; Lourenco, V.M.; Menezes, R.; Brites, C. Rice compounds with impact on diabetes control. *Foods* **2021**, *10*, 1992. [CrossRef]
30. AACC. AACC holds midyear meeting. *Cereal Food World* **2000**, *45*, 327.
31. Smith, C.E.; Tucker, K.L. Health benefits of cereal fibre: A review of clinical trials. *Nutr. Res. Rev.* **2011**, *24*, 118–131. [CrossRef]
32. Papandreou, D.; Noor, Z.T.; Rashed, M. The role of soluble, insoluble fibers and their bioactive compounds in cancer: A mini review. *Food Nutr. Sci.* **2015**, *6*, 1–11. [CrossRef]
33. Mudgil, D. The interaction between insoluble and soluble fiber. In *Dietary Fiber for the Prevention of Cardiovascular Disease*, 1st ed.; Samaan, R.A., Ed.; Academic Press: Oxford, UK, 2017; pp. 35–59. [CrossRef]
34. Pasakawee, K.; Laokuldilok, T.; Srichairatanakool, S.; Utama-ang, N. Relationship among starch digestibility, antioxidant, and physicochemical properties of several rice varieties using principle component analysis. *Curr. Appl. Sci. Technol.* **2018**, *18*, 133–144. [CrossRef]
35. Karupaiah, T.; Aik, C.K.; Heen, T.C.; Subramaniam, S.; Bhuiyan, A.R.; Fasahat, P.; Zain, A.M.; Wickneswari, R. A transgressive brown rice mediates favourable glycaemic and insulin responses. *J. Sci. Food Agric.* **2011**, *91*, 1951–1956. [CrossRef] [PubMed]
36. Prasad, V.S.S.; Hymavathy, P.; Babu, V.R.; Longvah, T. Nutritional composition in relation to glycemic potential of popular Indian rice varieties. *Food Chem.* **2018**, *238*, 29–34. [CrossRef]
37. Sasaki, S. Rice and prevention of type 2 diabetes: Narrative review of epidemiologic evidence. *J. Nutr. Sci. Vitamiol.* **2019**, *65*, S38–S41. [CrossRef]
38. Jenkins, D.J.; Jenskins, A.L.; Wolever, T.M.; Rao, A.V.; Thompson, L.U. Fiber and starchy foods: Gut function and implications in disease. *Am. J. Gastroenterol.* **1986**, *81*, 920–930.
39. Kondo, K.; Morino, K.; Nishio, Y.; Ishikado, A.; Arima, H.; Nakao, K.; Nakagawa, F.; Nikami, F.; Sekine, O.; Nemoto, K.I.; et al. Fiber-rich diet with brown rice improves endothelial function in type2 diabetes mellitus: A randomized controlled trial. *PLoS ONE* **2017**, *12*, e0189869. [CrossRef]
40. Sharma, P.; Bhandari, C.; Kumar, S.; Sharme, B.; Bhadwal, P.; Agnihotri, N. Dietary fibers: A way to a healthy microbiome. In *Diet, Microbiome and Heath: Handbook of Food Bioengineering Volume 11*, 1st ed.; Holban, A.M., Grumezescu, A.M., Eds.; Academic Press: Oxford, UK, 2018; pp. 299–345. [CrossRef]
41. Prasadi, N.; Joye, I.J. Dietary fibre from whole grains and their benefits on metabolic health. *Nutrients* **2020**, *12*, 3045. [CrossRef]

42. Vitaglione, P.; Napolitano, A.; Fogliano, V. Cereal dietary fibre: A natural functional ingredient to deliver phenolic compounds into the gut. *Trends Food Sci. Technol.* **2008**, *19*, 451–463. [CrossRef]
43. Lattimer, J.M.; Haub, M.D. Effects of dietary fiber and its components on metabolic health. *Nutrients* **2010**, *2*, 1266–1289. [CrossRef]
44. Gujral, H.S.; Kumar, V. Effect of accelerated aging on the physicochemical and textural properties of brown and milled rice. *J. Food Eng.* **2003**, *59*, 117–121. [CrossRef]
45. Parween, S.; Anonuevo, J.J.; Butardo, V.M., Jr.; Misra, G.; Anacleto, R.; Llorente, C.; Kosik, O.; Romero, M.V.; Bandonill, E.H.; Mendioro, M.S.; et al. Balancing the double-edged sword effect of increased resistant starch content and its impact on rice texture: Its genetics and molecular physiological mechanisms. *Plant Biotechnol. J.* **2020**, *18*, 1763–1777. [CrossRef] [PubMed]
46. Lu, S.; Cik, T.; Lii, C.; Lai, P.; Chen, H. Effect of amylose content on structure, texture and–amylose reactivity of cooked rice. *Food Sci. Tech.* **2013**, *54*, 224–228. [CrossRef]
47. Liu, Q.; Tao, Y.; Cheng, S.; Zhou, L.; Tian, J.; Xing, Z.; Liu, G.; Wei, H.; Zhang, H. Relating amylose and protein contents to eating quality in 105 varieties of *Japonica* rice. *Cereal Chem.* **2020**, *97*, 1303–1312. [CrossRef]
48. Mestres, C.; Ribeyre, F.; Pons, B.; Fallet, V.; Matencio, F. Sensory texture of cooked rice is rather linked to chemical than to physical characteristics of raw grain. *J. Cereal Sci.* **2011**, *53*, 81–89. [CrossRef]
49. Patindol, J.; Gu, X.; Wang, Y.J. Chemometric analysis of cooked rice texture in relation to starch fine structure and leaching characteristics. *Starch* **2010**, *62*, 188–197. [CrossRef]
50. Li, H.; Prakash, S.; Nicholson, T.M.; Fitzgerald, M.A.; Gilber, R.G. The importance of amylose and amylopectin fine structure for textural properties of cooked rice grains. *Food Chem.* **2016**, *196*, 702–711. [CrossRef]
51. Li, H.; Fitzgerald, M.; Prakash, S.; Nicholson, T.; Gilbert, R.G. The molecular structural features controlling stickiness in cookes rice, a major palatability determinant. *Sci. Rep.* **2017**, *7*, 43713. [CrossRef]
52. Ritika, B.Y.; Satnam, M.; Baljeet, S.Y. Physicochemical, pasting, cooking and textural quality characteristics of some basmati and non-basmati rice varieties grown in India. *Int. J. Agri. Technol.* **2016**, *12*, 675–692.
53. Bhat, F.M.; Riar, C.S. Physicochemical, cooking, and textural characteristics of grains of different rice (*Oryza sativa* L.) cultivars of temperate region of India and their interrelationships. *J. Texture Stud.* **2017**, *48*, 160–170. [CrossRef]
54. Chen, F.; Yang, C.; Liu, L.; Liu, T.; Wang, Y.; Wang, L.; Shi, Y.; Campanella, O.H. Differences, correlation of composition, taste and texture characteristics of rice from Heilongjiang China. *J. Rice Res.* **2017**, *5*, 1000178. [CrossRef]
55. Cuevas, R.P.O.; Domingo, C.J.; Sreenivasulu, N. Multivariate-based classification of predicting cooking quality ideotypes in rice (*Oryza sativa* L.) indica germplasm. *Rice* **2018**, *11*, 56-d69. [CrossRef] [PubMed]
56. Daou, C.; Zhang, H. Functional and physiological properties of total, soluble, and insoluble dietary fibres derived from defatted rice bran. *J. Food Sci. Technol.* **2014**, *51*, 3878–3885. [CrossRef]
57. Tao, K.; Yu, W.; Prakash, S.; Gilbert, R.G. Investigating cooked rice textural properties by instrumental measurements. *Food Sci. Hum. Wellness* **2020**, *9*, 130–135. [CrossRef]
58. Prosky, L.; Asp, N.G.; Schweizer, T.F.; DeVries, J.W.; Furda, I. Determination of insoluble and soluble dietary fiber in foods and food products: Collaborative study. *J. AOAC Int.* **1992**, *75*, 360–367. [CrossRef]
59. McCleary, B.V.; DeVries, J.W.; Rader, J.I.; Cohen, G.; Prosky, L.; Mugford, D.C.; Champ, M.; Okuma, K. Determination of Total Dietary Fiber (CODEX Definition) by Enzymatic-Gravimetric Method and Liquid Chromatography: Collaborative Study. *J. AOAC Int.* **2010**, *93*, 221–233. [CrossRef] [PubMed]
60. Juliano, B.O. Rice starch properties and grain quality. *Denpun Kagaku* **1992**, *39*, 11–21. [CrossRef]
61. Pang, Y.; Ali, J.; Wang, W.; Franje, N.J.; Revilleza, J.E.; Xu, J.; Li, Z. Relationship of Rice Grain Amylose, Gelatinization Temperature and Pasting Properties for Breeding Better Eating and Cooking Quality of Rice Varieties. *PLoS ONE* **2016**, *11*, e0168483. [CrossRef]
62. Liu, X.; Ding, Q.; Wang, W.; Pan, Y.; Tan, C.; Qiu, Y.; Chen, Y.; Li, H.; Li, Y.; Ye, N.; et al. Targeted deletion of the first intron of the Wx^b Allele via CRISPR/Cas9 significantly increases grain amylose content in rice. *Rice* **2022**, *15*, 1. [CrossRef]
63. Gujral, H.S.; Singh, J.; Sodhi, N.S.; Singh, N. Effect of milling variables on the degree of milling of unparboiled and parboiled rice. *Int. J. Food Prop.* **2002**, *5*, 193–204. [CrossRef]
64. Bautista, R.C.; Siebenmorgen, T.J. Evaluation of laboratory mills for milling small samples of rice. *Appl. Eng. Agric.* **2002**, *18*, 577–583. [CrossRef]
65. Little, R.R.; Hilder, G.B.; Dawson, E.H. Differential effect of dilute alkali on 25 varieties of milled white rice. *Cereal Chem.* **1958**, *35*, 111–126.
66. Bhattacharya, K.R.; Sowbhagya, C.M. An improved alkali reaction test for rice quality. *J. Food Technol.* **1972**, *7*, 323–331. [CrossRef]
67. Ren, X.; Zhao, X.; Turcotte, F.; Deschênes, J.S.; Tremblay, R.; Jolicoeur, M. Current lipid extraction methods are significantly enhanced adding a water treatment step in Chlorella protothecoides. *Microb. Cell Fact.* **2017**, *16*, 26. [CrossRef]
68. Ohkuma, K.; Matsuda, I.; Katta, Y. New method for determining total dietary fiber by liquid chromatography. *J. AOAC Int.* **2000**, *83*, 1013–1019. [CrossRef]
69. Nádvorníková, M.; Banout, J.; Herák, D.; Verner, V. Evaluation of physical properties of rice used in traditional Kyrgyz cuisine. *Food Sci. Nutr.* **2018**, *6*, 1778–1787. [CrossRef]
70. Chen, M.H.; McClung, A.M. Genotypic diversity of bran weight of whole grain rice and its relationship with grain physical traits. *Cereal Chem.* **2018**, *96*, 252–262. [CrossRef]
71. Juliano, B.O. A simplified assay for milled-rice amylose. *Cereal Sci. Today* **1971**, *16*, 335–340.

72. Guillen, S.; Oria, R.; Salvador, M.L. Impact of Cooking Temperature on In Vitro Starch Digestibility of Rice Varieties with Different Amylose Contents. *Pol. J. Food Nutr. Sci.* **2018**, *68*, 319–325. [CrossRef]
73. Bernstein, I.H. Likert Scale Analysis. In *Encyclopedia of Social Measurement*, 1st ed.; Leonard, K.K., Ed.; Elsevier Inc.: Amsterdom, The Netherlands, 2005; pp. 497–504.
74. Statistic Solutions. Available online: https://www.statisticssolutions.com (accessed on 16 September 2022).
75. The R Project for Statistical Computing. Available online: https://www.r-project.org (accessed on 16 September 2022).
76. Lyon, B.G.; Champagne, E.T.; Vinyard, B.T.; Windham, W.R. Sensory and instrumental relationships of texture of cooked rice from selected cultivars and postharvest handling practices. *Cereal Chem.* **2000**, *77*, 64–69. [CrossRef]
77. Kohyama, K. Food texture—Sensory evaluation and instrumental measurement. In *Textural Characteristics of World Foods*, 1st ed.; Nishinari, K., Ed.; John Wiley & Sons, Inc.: Hoboken, NJ, USA, 2020; pp. 1–13. [CrossRef]
78. Cuevas, R.P.O.; Takhar, P.S.; Sreenivasulu, N. Characterization of mechanical texture attributes of cooked milled rice by texture profile analyses and unraveling viscoelasticity properties through rheometry. *Methods Mol. Biol.* **2019**, *1892*, 151–167. [CrossRef]
79. Lim, J. Hedonic scaling: A review of methods and theory. *Food Qual. Pref.* **2011**, *22*, 733–747. [CrossRef]
80. Abeysekera, W.K.S.M.; Arachchige, S.P.G.; Ratnasooriya, W.D.; Chandrasekharan, N.V.; Bentota, A.P. Physicochemical and nutritional properties of twenty three traditional rice (*Oryza sativa* L.) varieties of Sri Lanka. *J. Coast. Life Med.* **2017**, *5*, 343–349. [CrossRef]
81. Hamid, A.A.; Luan, Y.S. Functional properties of dietary fiber from defatted rice bran. *Food Chem.* **2000**, *68*, 15–19. [CrossRef]
82. Faria, S.A.S.C.; Bassinello, P.Z.M.; Penteado, V.C. Nutritional composition of rice bran submitted to different stabilization procedures. *Braz. J. Pharm. Sci.* **2012**, *48*, 651–657. [CrossRef]
83. Zhang, C.; Zhu, J.; Chen, S.; Fan, X.; Li, Q.; Lu, L.; Wang, M.; Yu, H.; Yi, C.; Tang, S.; et al. Wx^{lv}, the Ancestral Allele of Rice Waxy Gene. *Mol. Plant* **2019**, *12*, 1157–1166. [CrossRef]
84. Champagne, E.T. *Rice: Chemistry and Technology*, 3rd ed.; AACC International Press: Washington, DC, USA, 2004; pp. 77–190. ISBN 978-1-891127-34-2.
85. Juliano, B.O.; Tuaño, A.P.P. Gross structure and composition of the rice grain. In *Rice, Chemistry and Technology*, 4th ed.; Bao, J., Ed.; Woodhead Publishing: Cambridge, UK, 2019; pp. 31–53. [CrossRef]
86. Din, S.U.; Bhattacharya, K.R. On the meaning of the degree of milling of rice. *J. Food Technol.* **1978**, *13*, 99–105. [CrossRef]
87. Pokhrel, A.; Dhakal, A.; Sharma, S.; Poudel, A. Evaluation of physicochemical and cooking characteristics of rice (*Oryza sativa* L.) Landraces of Lamjung and Tanahun districts, Nepal. *Int. J. Food Sci.* **2020**, *2020*, 1589150. [CrossRef]
88. Bett-Garber, K.L.; Lea, J.M.; McClung, A.M.; Chen, M.H. Correlation of sensory, cooking, physical, and chemical properties of whole grain rice with diverse bran color. *Cereal Chem.* **2013**, *90*, 521–528. [CrossRef]
89. Mariotti, M.; Fongaro, L.; Catenacci, F. Alkali spreading value and image analysis. *J. Cereal Sci.* **2010**, *52*, 227–235. [CrossRef]
90. Tuaño, A.P.P.; Ricafort, C.H.; del Rosario, E.J. Estimation of alkali spreading value and gelatinization temperature of some Philippine rice varieties using digital photometer. *Philipp. Agric. Sci.* **2018**, *101*, 354–362.
91. Lim, S.T.; Lee, J.H.; Shin, D.H.; Lim, H.S. Comparison of protein extraction solutions for rice starch, isolation and effects of residual protein content on starch pasting properties. *Starch* **1999**, *4*, 120–125. [CrossRef]
92. Phongthai, S.; Homthawornchoo, W.; Rawdkuen, S. Preparation, properties and application of rice bran protein: A review. *Int. Food Res. J.* **2017**, *24*, 25–34.
93. Bender, D.; Nemeth, R.; Wimmer, M.; Gotschhofer, S.; Biolchi, M.; Torok, K.; Tomoskozi, S.; D'Amico, S.; Schoenlechner, R. Optimization of arabinoxylan isolation from rye bran by adapting extraction solvent and use of enzymes. *J. Food Sci.* **2017**, *82*, 2562–2568. [CrossRef]
94. Zhang, H.; Wu, F.; Xu, D.; Xu, X. Endogenous alpha-amylase explains the different pasting and rheological properties of wet and dry milled glutinous rice flour. *Food Hydrocoll.* **2021**, *113*, 106425. [CrossRef]
95. Yasui, T.; Sasaki, T.; Matsuki, J. Variation in the Chain-length Distribution Profiles of Endosperm Starch from Triticum and Aegilops Species. *Starch* **2005**, *57*, 521–530. [CrossRef]
96. Englyst, H.N.; Quigley, M.E.; Hudson, G.J. Dietary fiber analysis as non-starch polysaccharides. In *CRC Handbook of Dietary Fiber in Human Nutrition*; CRC Press: Boca Raton, FL, USA, 2006. [CrossRef]
97. Butardo, V.M.; Sreenivasulu, N. Tailoring grain storage reserves for a healthier rice diet and its comparative status with other cereals. *Int. Rev. Cell Mol. Biol.* **2016**, *323*, 31–70. [CrossRef]
98. Lovergrove, A.; Kosik, O.; Bandonill, E.; Abilgos-Ramos, R.; Romero, M.; Sreenivasulu, N.; Shewry, P. Improving rice dietary fiber content and composition for human health. *J. Nutr. Sci. Vitaminol.* **2019**, *65*, S48–S50. [CrossRef]
99. Kosik, O.; Romero, M.V.; Bandonill, E.H.; Abilgos-Romos, R.G.; Sreenivasulu, N.; Shewry, P.; Lovegrove, A. Diversity of content and composition of cell wall-derived dietary fibre in polished rice. *J. Cereal Sci.* **2020**, *96*, 10312. [CrossRef]
100. Lai, V.M.F.; Lu, S.; He, W.H.; Chen, H.H. Non-starch polysaccharide compositions of rice grains with respect to rice variety and degree of milling. *Food Chem.* **2006**, *101*, 1205–1210. [CrossRef]
101. Dodevska, M.S.; Djordjevic, B.I.; Sobajic, S.S.; Miletic, I.D.; Djordjevic, P.B.; Dimitrijevic-Sreckovic, V.S. Characterisation of dietary fibre components in cereals and legumes used in Serbian diet cereals and legumes used in Serbian diet. *Food Chem.* **2013**, *141*, 1624–1629. [CrossRef] [PubMed]
102. Kumar, V.; Sinha, A.; Makkar, H.P.S.; de Boeck, G.; Becker, K. Dietary roles of non-starch polysaccharides in human nutrition: A Review. *Crit. Rev. Food Sci. Nutr.* **2012**, *52*, 899–935. [CrossRef] [PubMed]

103. BeMiller, J. Pectins. In *Carbohydrate Chemistry for Food Scientists*, 3rd ed.; BeMiller, J., Ed.; Workhead Publishing and AACC International Press: Cambridge, UK, 2018; pp. 303–312. [CrossRef]
104. Bhattacharya, K.R. Cooking quality of rice. In *Rice Quality: A Guide to Rice Properties and Analysis*, 1st ed.; Bhattacharya, K.R., Ed.; Woodhead Publishing: Cambridge, UK, 2011; pp. 164–192. [CrossRef]
105. Ahmed, F.; Abro, T.F.; Kabir, M.S.; Latif, M.A. Rice quality: Biochemical composition, eating quality, and cooking quality. In *The Future of Rice Demand: Quality Beyond Productivity*, 1st ed.; De Oliveira, A.C., Pegoraro, C., Viana, V.E., Eds.; Springer: Cham, Switzerland, 2020; pp. 3–24. [CrossRef]
106. Dhingra, D.; Michael, M.; Rajput, H.; Patil, R.T. Dietary fibre in foods: A review. *J. Food Sci. Technol.* **2012**, *49*, 255–266. [CrossRef]
107. El Khoury, D.; Cuda, C.; Luhovyy, B.L.; Anderson, G.H. Beta glucan: Health benefits in obesity and metabolic syndrome. *J. Nutr. Metab.* **2012**, *2012*, 851362. [CrossRef] [PubMed]
108. Slavin, J. Fiber and prebiotics: Mechanisms and health benefits. *Nutrients* **2013**, *5*, 1417–1435. [CrossRef] [PubMed]
109. Phuwadoipaisarn, P. The influence of conditions on beta-glucan extraction from Thai rice bran cultivars and their biological properties. In Proceedings of the 95th The IIER International Conference, Osaka, Japan, 8–9 February 2017; ISBN 978-93-86083-34-0.
110. Neethi, R.P.; Anie, Y. Extraction of water-soluble beta-glucan from rice bran. *Sch. Acad. J. Biosci.* **2017**, *5*, 766–770.

foods
MDPI

Article

Development of Gluten-Free Bread Using Teosinte (*Dioon mejiae*) Flour in Combination with High-Protein Brown Rice Flour and High-Protein White Rice Flour

Franklin Delarca Ruiz [1], Ricardo S. Aleman [2], Shirin Kazemzadeh Pournaki [3], Mallerly Sarmiento Madrid [1], Andrea Muela [2], Yeimi Mendoza [2], Jhunior Marcia Fuentes [1], Witoon Prinyawiwatkul [2] and Joan M. King [2,*]

1 Faculty of Technological Sciences, Universidad Nacional de Agricultura, Catacamas 16201, Honduras; fdelarca19a0218@unag.edu.hn (F.D.R.); msarmiento19a0164@unag.edu.hn (M.S.M.); jmarcia@unag.edu.hn (J.M.F.)
2 School of Nutrition and Food Sciences, Louisiana State University Agricultural Center, Baton Rouge, LA 70803, USA; rsantosaleman@lsu.edu (R.S.A.); amuela1@lsu.edu (A.M.); ymendoza@agcenter.lsu.edu (Y.M.); wprinyawiwatkul@agcenter.lsu.edu (W.P.)
3 Department of Dairy & Food Science, South Dakota State University, Brookings, SD 57007, USA; shirin.kazemzadehpournaki@sdstate.edu
* Correspondence: jking@agcenter.lsu.edu

Abstract: Gluten-free bread is an important product that is under development using different sources, such as rice and starchy plants. Teosinte seeds are utilized by ethnic groups in Honduras to produce gluten-free flour to prepare traditional baked goods and beverages. The quality of gluten-free products could vary depending on flour properties, such as amylose content, particle size, and water absorption capacity. A good strategy for developing baked goods is to mix different cereal grain sources to optimize their physicochemical properties. As a result, the current study aimed to develop bread from novel flours including teosinte (TF), high-protein brown rice (BRF), and high-protein white rice (WRF). Breads were analyzed for hardness, specific volume, and color utilizing a Simplex-Centroid mixture design coupled with the desirability function. Pasting, and rheological characteristics of the flours, were also analyzed. For flour characteristics, TF addition to BRF or WRF decreased the peak, trough, breakdown, setback, and final viscosities, which would result in a more stable bread and decrease the flow index of rice flour dispersions. BRF and WRF had similar pasting properties, except that BRF had a lower breakdown viscosity. For bread characteristics, TF addition to BRF or WRF increased the specific volume and hardness of the bread compared to rice flour alone. L* of the crust and crumb a* values were increased with greater TF in the mixture, whereas TF decreased the crust a*and b* values and crumb L* values when mixed with BRF or WRF compared to rice flours alone. WRF and BRF were similar in crumb color (L* and a*), except that BRF had greater crumb yellowness (b*). Teosinte flour can be used in combination with rice flour to produce bread with good quality.

Keywords: teosinte; rice; gluten free; bread; mixture design

Citation: Delarca Ruiz, F.; Aleman, R.S.; Kazemzadeh Pournaki, S.; Sarmiento Madrid, M.; Muela, A.; Mendoza, Y.; Marcia Fuentes, J.; Prinyawiwatkul, W.; King, J.M. Development of Gluten-Free Bread Using Teosinte (*Dioon mejiae*) Flour in Combination with High-Protein Brown Rice Flour and High-Protein White Rice Flour. *Foods* **2023**, *12*, 2132. https://doi.org/10.3390/foods12112132

Academic Editors: Jennifer Ahn-Jarvis and Brittany A. Hazard

Received: 3 May 2023
Revised: 22 May 2023
Accepted: 24 May 2023
Published: 25 May 2023

1. Introduction

Bread uses the ability of hydrated gluten to build a viscoelastic network [1], which causes gas to be trapped, and thus increases its volume. Additionally, gluten in bread plays a key role in moisture control [2]. It is a high-molecular weight protein found in the endosperm of cereals including wheat, barley, and rye. Additionally, it is a storage protein in a group of flowering plants utilized during the growth and germination process that consists of two types of proteins, i.e., a glutenin and a prolamin (gliadin found in wheat), which can be broken down to produce α, β, and γ peptides [3].

However, gluten in bread can cause problems in a group of consumers. Celiac disease is an autoimmune disorder seen in people who genetically have the potential to develop an immune reaction to gluten. The first place affected by this disease is the small intestine; nevertheless, it has a wide distribution that includes both intestinal and extra-intestinal symptoms [4]. Currently, the most effective and safest treatment for people affected by Celiac disease is to use a gluten-free (GF) diet, which causes an improvement of the small intestinal mucosa [5,6].

Gluten helps by forming a sticky, elastic dough with gas retention, as well as shaping the structure [7]. In the absence of gluten in bread, the ability of the bread to hold carbon dioxide produced by yeast is significantly reduced, which results in bread with a firm texture, as well as low specific volume [8].

Gluten-free bread available in the market is usually obtained by replacing wheat flour with rice flour with or without corn starch. Rice flour is known for its low price and mild taste, as well as antiallergenic properties, which is why it is utilized in these products. The rice flour used to prepare gluten-free bread is mostly from white rice [9]. However, gluten-free bread, which is obtained by using white rice flour, is nutritionally imbalanced due to the removal of the entire bran layer, leaving mainly starch and protein. It is expected that the use of brown rice flour in gluten-free bread will compensate for this deficiency because brown rice has non-starch nutrients including dietary fiber, minerals, and bioactive compounds in its bran layer [10].

Teosinte (*Dioon mejiae*) as an endemic tree in Honduras is one of the dioecious trees and belongs to the category of minor cereals. Minor cereals occur only in a few parts of the world, that is why their use on a large scale is not common. In addition to Teosinte, other plants including teff, millet (pearl, proso, finger, foxtail, and Kodo), fonio (black and white), jungle rice, and Job's tears are in this category [11–14]. Teosinte seeds are locally used in the preparation of flour as well as other traditional foods and drinks. The nutritional value of teosinte seed is high with protein and methionine levels higher than maize. However, no difference was reported in the amino acids, such as lysine, tryptophan, or niacin [15]. In the distant past, regional people used the seeds of this plant after washing and drying to prepare foods such as bread, donuts, tamales, and tortillas. Additionally, a type of starch called sago is obtained from this plant, which is used as a food supplement by the natives of that area [16].

Some studies have been conducted in the field of gluten-free bread production and improvement. Some innovative technologies were proposed to improve quality, replace or imitate the gluten network by using exogenous substances including hydrocolloids, emulsifiers, proteins, and cross-linking enzymes [6,17]. As mentioned previously, among the available approaches for gluten-free bread production is the use of alternative plant sources for wheat. Different types of flour and starch (rice, corn, cassava, soybean, and peanut) have been used to produce gluten-free bread [18,19]. For example, active soybean flour improved the volume and structure of gluten-free bread [19]. Bread made with corn flour and chickpea flour became softer with greater levels of chickpea flour, which was thought to be due to greater protein levels [20]. Other flours are used to increase the nutritional quality of gluten free breads [21].

In this regard, due to the essential need to diversify the diet and for those who have special nutritional requirements, such as celiac patients, it is mandatory to provide an innovative diet. For this purpose, examining the potential of traditional and lesser-known food sources as alternatives and the expansion of their use to produce innovative gluten-free foods that are acceptable to consumers seems promising. Due to the potential for using brown rice flour and teosinte flour in gluten-free foods, and the few studies in this field, this study aims to use these plant resources to develop gluten-free bread.

2. Material and Method

2.1. Experimental Design

Table 1 shows the treatment design for bread production using a Simplex-Centroid Mixture Design (CSMD). The independent variables were the proportions of high-protein brown rice flour (Cahokia), high-protein white rice flour (Cahokia), and teosinte flour, while the dependent variables included specific volume (g/mL), color (L*, b* and a*) and texture (resilience, cohesiveness, hardness, and springiness). The obtained response from each investigated parameter was analyzed using adjusting the cubic model (Y = $\beta1 \times 1 + \beta2 \times 2 + \beta3 \times 3 + \beta12 \times 1 \times 2 + \beta13 \times 1 \times 3 + \beta23 \times 2 \times 3 + \beta123 \times 1 \times 2 \times 3$) at $p < 0.05$ and using regression to determine significant differences in parameters for the level of flour/starch used (dependent variables). The bread formula was optimized using the desirability function methodology (DOM) [22]. The objective of the mixture design is to optimize flour concentrations of high-protein brown rice flour (Cahokia), high-protein white rice flour (Cahokia), and teosinte flour regarding physicochemical characteristics.

Table 1. Experimental design for bread making with simplex centroid design.

Treatments	*TF	*BRF	*WRF
▪ TF alone	100%	0%	0%
▪ BRF alone	0%	100%	0%
▪ WRF alone	0%	0%	100%
▪ TF-BRF	50%	50%	0%
▪ TF-WRF	50%	0%	50%
▪ BRF-WRF	0%	50%	50%
▪ TF-BRF-WRF	33.337%	33.337%	33.337%

*BRF—High-Protein Brown Rice Flour, *TF—Teosinte Flour, *WRF—High-Protein White Rice.

2.2. Preparing the Bread

The amounts of each ingredient used are shown in Table 2 for gluten-free breads and control breads (the same procedure was used for both type of breads and only the formulation varied). To make the bread, Fleischmann's activated dry yeast (ACH Food Companies, Inc., Memphis, TN, USA) was mixed with sugar (Great Value, Leander, TX, USA) and distilled water, then rested for 42 h. Subsequently, by using a Globe stand mixer (model SP5 Global Food Equipment, Dayton, OH, USA), the other dry ingredients (Great Value, Leander, TX, USA) were mixed. The mixture was gently stirred for 30 s. Next, in a separate container, room temperature eggs (Great Value, Leander, TX, USA), vegetable oil (Great Value, Leander, TX, USA), and apple cider vinegar (Great Value, Leander, TX, USA) were weighed and added to the dry ingredients and gently mixed for 1 min. Then, the yeast mixture was gradually added and mixed for 7 min. In the production process of gluten-free bread, the consistency is usually similar to batter (instead of dough, which can be kneaded). Vegetable oil was sprayed in a mini loaf pan (15.4 × 8.6 × 4.7 cm) to grease. Next 150 g of standard-loaf batter was weighed in the pan and the surface was smoothed with a spatula. The relative humidity and temperature of a full-size Metro proofing cabinet (C599-SDS-U Intermetro Industries Corporation, Wilkes-Barre, PA, USA) were set to 90% and 100 °F, respectively, then the pan was placed in the cabinet for 30 min. Afterwards the pan was placed in the center of a Baxter mini-rotating rack gas oven (model OV310G) at a temperature of 345 °F and baked for 20 min. At the end of the baking process, the bread was left in the pan for 5 min to cool and then removed from the pan. After an hour of cooling, a sanitized, electric, meat-slicing machine (model S-4 Sanitary Scale Company, Belvidere, IL, USA) was used to prepare slices of 2.5 cm for color and texture analysis.

Table 2. Percentages of gluten-free bread ingredients.

Gluten Free Breads		Control Bread	
Ingredients	**Percentage**	**Ingredient**	**Percentage**
Flour *	17.09%	All purpose flour	42.32%
Tapioca flour	14.64%	Whole wheat flour	10.83%
Sugar	3.33%	Sugar	6.66%
Salt	0.98%	Salt	1.50%
Active dry yeast	0.88%	Active dry yeast	1.17%
Water	32.2%	Water	30.67%
Vegetable oil	1.47%	Vegetable oil	4.89%
Vanilla	1.96%	Vanilla	1.96%
Cornstarch	16.11%		
Egg	9.76%		
Xanthan gum	0.78%		
Baking powder	0.49%		
Apple cider vinegar	0.29%		

* Treatments varied only by mixture design treatments illustrated in Table 1. Control bread = wheat bread.

2.3. Flour Rheological and Pasting Properties

The pasting properties of the flours were evaluated based on the AACC method 61.02.01 [23] by using a Rapid Visco Analyzer (RVA) (RVA-4, Newport Scientific Pty. Ltd., Warriewood, Australia). Rheological properties were evaluated with a rheometer (AR 2000ex, TA Instruments, New Castle, DE, USA) by parallel disc geometry and 40 mm dimensions with a gap of 3 mm. Dispersions of 5% *w*/*w* were stirred at medium speed for 15 min and heated for 30 min. Instantly, hot paste (1 mL) was placed in the rheometer. When the sample temperature reached 25 °C, rheological analysis was performed with two types of evaluation (steady shear flow as well frequency sweep) using the method from Ye et al. [24] with slight modifications.

2.4. Bread Physical Features

The specific volume of bread (mL/g) was measured according to the AACC method 10-05 [25] with rapeseed. The texture of the sample was analyzed with a texture analyzer (Texture Technologies Corporation, T.A. XT plus, Scarsdale, NY, USA) based on the AACC method 74-09 (2000) using a two-inch cylinder probe. The bread was cut into to 2.5 cm width slices to examine the bread's firmness. The parameters were set to a 40% compression at a rate of 1.7 mm/s. With colorimeter equipment (Konica Minolta BC-10 Baking Contrast Meter, Wayne, NJ, USA), L* (brightness/darkness), a* (redness/greenness), and b* (yellow/bluish) samples were analyzed in triplicate.

2.5. Statistical Analysis

For the simplex-centroid design, generation of the corresponding response surfaces and coefficients of the special cubic model was performed in the Minitab 17 program (2014, Minitab LLC, State College, PA, USA) to check the characteristics of bread. One-way analysis of variance (ANOVA) and Tukey's post-hoc test were conducted.

3. Results and Discussion

3.1. Bread Characteristics

Prepared bread samples are shown in Figure 1 compared to control wheat flour bread.

Figure 2A shows that the bread sample prepared with a combination of TF*BRF presented a greater specific volume compared to BRF and TF individually. Kadan et al. [26] found that rice bread had a lower specific volume than wheat bread; in our study the addition of TF to BRF resulted in a greater specific volume. According to Bastias-Montes et al. [11], protein and total starch contents were 9.67 ± 0.08% and 67.90 ± 0.68% in teosinte flour, and Aleman et al. [27] found that high-protein brown rice flour had 12.2 ± 0.14% protein and 65.4 ± 0.5% starch, while high-protein white rice flour had 10.23 ± 0.26% protein and

75.15 ± 0.20% starch [28]. The specific volume is directly related to water absorption of the network, which affects bread quality, and bread with greater starch content tended to have a greater specific volume [29]. Our study does not show this, as bread with a greater protein, TF*BRF, had a greater specific volume, which could be due to the protein also binding to the water and stabilizing the starch gel [30]. Additionally, the mixture which contained equal percentages of TF, BRF, and WRF showed a lower specific volume similar to bread with WRF or BRF alone. Furthermore, TF alone had the greatest hardness (Figure 2B) compared to other treatments, and Table 3 coefficients indicated that TF had a greater positive effect on the hardness than other flours, which might be due its lower protein content which leads to less water binding and decreased the loaf specific volume [30]. Greater hardness is associated with a lower specific volume, as less water binding could lead to a dryer firmer product with less stable air pockets. WRF alone and BRF alone had the lowest hardness with WRF being lower than BRF. Paz et al. [31] found a similar result for hardness between brown and white rice flour breads.

Crust lightness (L*) was lowest for TF*WRF bread, while TF*WRF*BRF bread had a greater lightness compared to bread made from BRF*TF and BRF*WRF (Figure 2C). Crust redness was the lowest in 100% TF bread (Figure 2D). Bread with TF*WRF and TF*BRF had lower a* compared to BRF*WRF. In another study, the addition of kale to bread decreased a* value due to the addition of the green color [32]. WRF increased the redness in bread which was made from a blend of WRF and BRF. The crust yellowness pattern (Figure 2E) was the same as the redness pattern which means WRF resulted in a greater yellowness in WRF alone, BRF-WRF, and TF-BRF-WRF. TF resulted in a lower yellowness in bread crust for TF alone and TF-WRF.

The crumb color pattern was different from the crust pattern. The lightness of the bread crumb (Figure 2F) was lowest with the use of TF alone, which might be due to the natural, brownish pigments in the seeds [33]. WRF was responsible for lightness in WRF alone, TF-WRF, and BRF-WRF. TF alone had the highest a*, and T_3 had the lowest redness (Figure 2G) among breadcrumbs. According to Figure 2H, WRF tended to decrease yellowness (b*) in TF-WRF and BRF-WRF. BRF breadcrumb had the greatest yellowness compared to other treatments, which may be due to the presence of yellow pigments [34].

Figure 1. Appearance of central slices of breads crumbs. (T1 = TF = Teosinte Flour, T2 = BRF = High-Protein Brown Rice, T3 = WRF = High-Protein White Rice Flour, T4 = TRF-BRF = Teosinte Flour with High-Protein Brown Rice, T5 = TF-WRF = Teosinte Flour with High-Protein White Rice Flour, T6 = BRF-WRF = High-Protein White Rice Flour with High-Protein Brown Rice Flour, T7 = TF-BRF-WRF = Teosinte Flour with High-Protein White Rice Flour and with High-Protein Brown Rice Flour) Control (T8) = bread made with wheat flour.

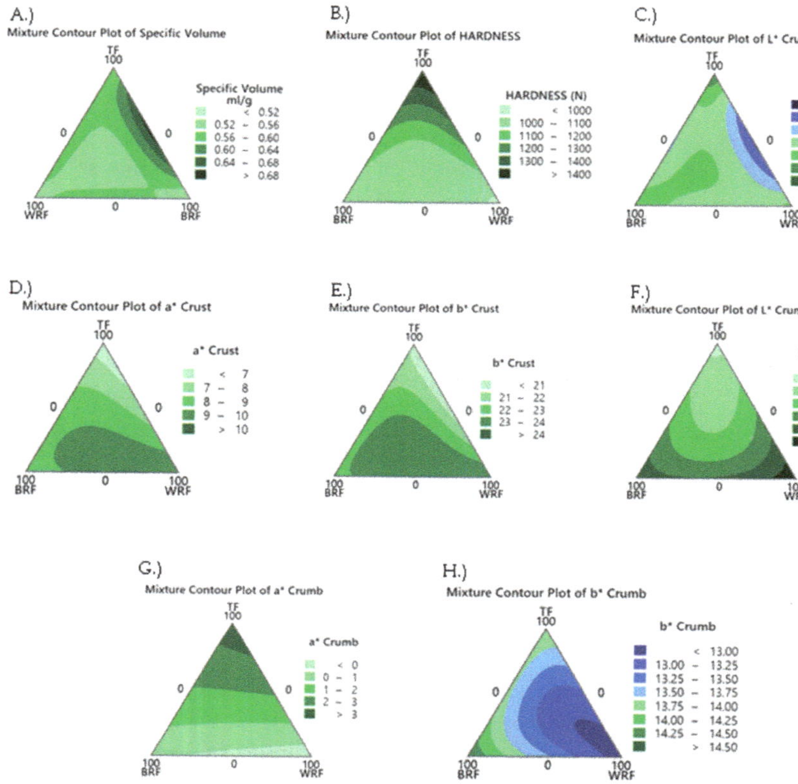

Figure 2. Contour plot of the breads (**A**) Specific Volume, (**B**) Hardness, (**C**) L* Crust, (**D**) a* Crust, (**E**) b* Crust, (**F**) L* Crumb, (**G**) a* Crumb, and (**H**) b* Crumb with *BRF—High-Protein Brown Rice Flour, *TF—Teosinte Flour, and *WRF—High-Protein White Rice.

3.2. Flour Pasting and Rheological Properties

Figure 3 shows peak, trough, breakdown, final, and setback viscosities, as well as yield stress, and flow behavior index with coefficients shown in Table 3. Figure 3A shows that TF alone had the lowest peak viscosity with the lowest coefficient (275) (Table 3) among the flours, while the addition of WRF to the bread blend increased the peak viscosity with a greater coefficient (1739.7), meaning a greater positive effect on peak viscosity (Table 3) than TF. The lower peak viscosity (Figure 3A) of TF resulted in the lowest breakdown viscosity (Figure 3C) coefficient (6.7) (Table 3), and greater past stability compared to BRF and WRF. BRF and WRF showed a greater positive influence on peak, trough final and setback viscosities with greater coefficients than TF (Table 3). Combinations of TF combined with BRF or WRF caused a negative influence resulting in lower pasting viscosities. Other model coefficients did not differ greatly among the individual flours (Table 3). Greater setback viscosity (Figure 2E) in BRF-WRF is associated with a greater retrogradation potential of the starch granules after cooking, which means it might cause a firmer product over time. However, there is no significant correlation between viscosity and starch concentration or distribution of the granules [35].

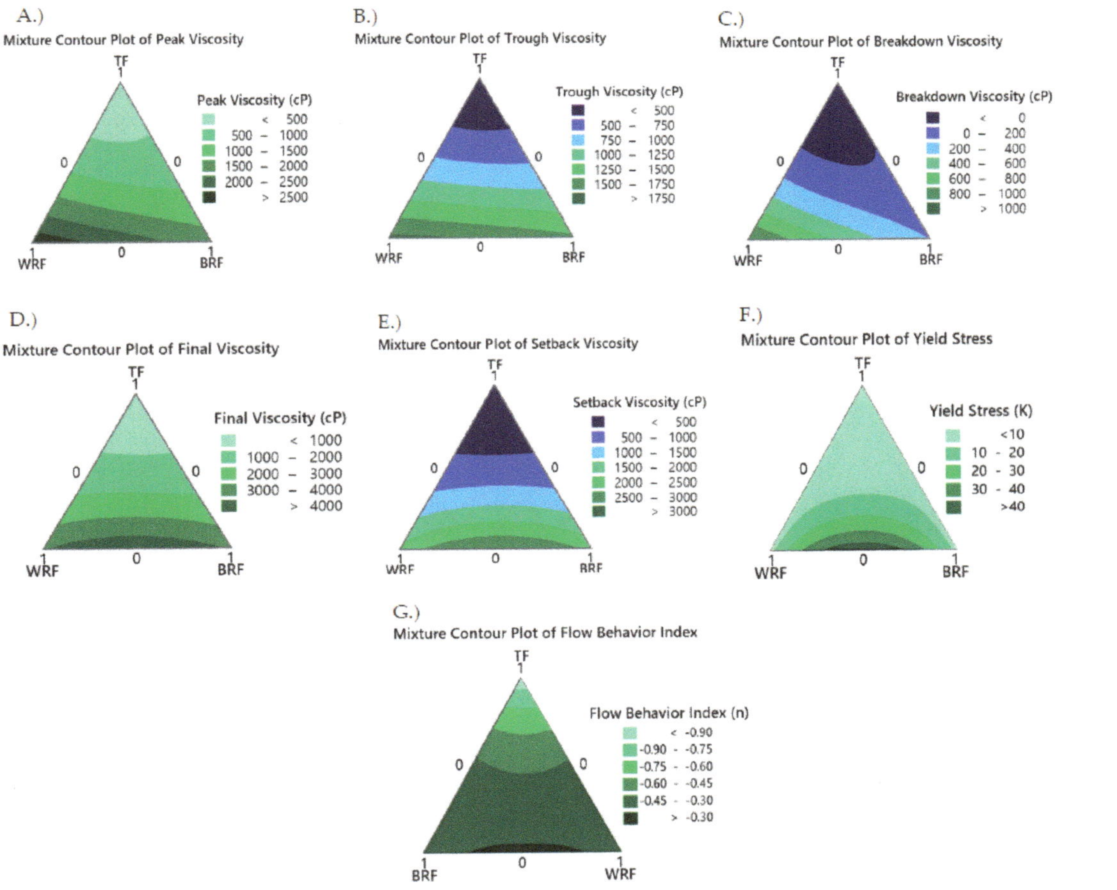

Figure 3. Contour plot of the flour dispersions (**A**) Peak Viscosity, (**B**) Trough Viscosity, (**C**) Breakdown Viscosity, (**D**) Final Viscosity, (**E**) Setback Viscosity, (**F**) Yield Stress, and (**G**) Flow Behavior Index with *BRF—High-Protein Brown Rice Flour, *TF—Teosinte Flour, and *WRF—High-Protein White Rice.

The combination of the WRF and BRF increased the yield stress and flow behavior index and TF had the lowest yield stress and flow behavior index (Figure 3F,G). According to the current study, all flow behavior indices were less than 1.0, indicating that all pastes exhibited pseudoplastic and shear-thinning behavior. WRF and BRF showed results closest to $n = 1$, which corresponds to a Newtonian fluid [36].

The complex viscosity (Figure 4A) decreases with increasing frequencies representing a shear-thinning flow behavior. The shear stress as a function of the shear rate is indicated in Figure 4B. which shows all treatments had non-Newtonian behavior due to the increase of the shear stress with the shear rate-like [37] pseudoplastic behavior. BRF-WRF had high-shear stress compared to other treatments. The pasting behavior of treatments is exhibited in Figure 4C, and various parameters were measured like peak, trough, breakdown, final, and setback viscosity. The shape of the pasting curve is different for different flour treatments and significant differences can be observed among GF flours, confirming the contour plot illustrations (Figure 3). The pasting curves of WRF alone had the highest value at peak viscosity and WRF-BRF had the highest final viscosity. The pasting process is the absorption of water by starch granules and granules lose their crystalline structure after swelling properly. According to the steady peak of TF for pasting over time and the heating process, amylose double helices were not melted in the cooking process, granules resisted swelling [38], and granules had a lower rate of water absorption and swelling compared to other samples [39]. According to Table 3, a greater influence was observed for BRF on peak viscosity, as well as a trough, breakdown, and final viscosity, than other single flour samples, while TF had a negative effect on these parameters when mixed with BRF. A higher protein could lead to lower peak viscosity, affecting peak time, trough, and breakdown viscosities by lowering the water-holding capacity of the starch during gelatinization [40]. Peak viscosity, trough viscosity, and breakdown viscosity for WRF were greater than BRF, which may be due to the greater starch content in WRF, 75.15 ± 0.20% vs. 65.4 ± 0.5% for BRF [37].

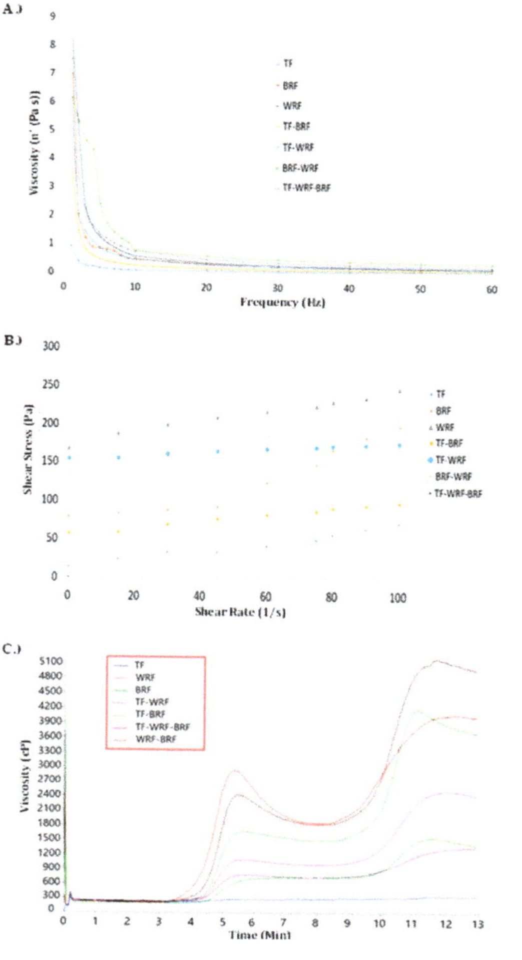

Figure 4. (**A**) steady shear flow measurements, (**B**) frequency sweep analysis, and (**C**) Viscosity profiles of flour dispersions (TF = Teosinte Flour, BR = High-Protein Brown Rice, WF = High-Protein White Rice Flour, TF-BF = Teosinte Flour with High-Protein Brown Rice, TF-WF = Teosinte Flour with High-Protein White Rice Flour, BF-WF = High-Protein White Rice Flour with High-Protein Brown Rice Flour, TF-BF-WR = Teosinte Flour with High-Protein White Rice Flour and with High-Protein Brown Rice Flour).

Table 3. Coefficients for the cubic model for bread and flours characteristics.

Response	TF	BRF	WRF	TF-BRF	TF-WRF	BRF-WRF	TF-BRF-WRF	R^2
Specific Volume (mL/g)	0.01	0.01	0.01	>0.01	>0.01	>0.01	>0.01	87.0%
Hardness (N)	14.9	10.5	9.68	−0.03	0.01	0.03	−0.03	92.8%
L* (Crust)	0.74	0.73	0.72	>−0.01	>−0.01	>−0.01	>0.01	80.9%
a∗ (Crust)	0.06	0.08	0.10	0.01	>−0.01	>0.01	>0.01	95.6%
b∗ (Crust)	0.20	0.22	0.24	>0.01	>−0.01	>0.01	>0.01	87.7%
L* (Crumb)	0.69	0.77	0.79	>−0.01	>−0.01	>−0.01	>−0.01	92.7%
a* (Crumb)	0.03	>0.01	>−0.01	>0.01	>0.01	>−0.01	>−0.01	99.4%
b* (Crumb)	0.14	0.15	0.12	>−0.01	>−0.01	>−0.01	>−0.01	75.5%
Peak Viscosity (cP)	275	2848.3	1739.7	−2909	−1164	−29	−2533	99.2%
Trough Viscosity (cP)	268.3	1811.7	1541.3	−1128	−814	181	−1253	98.6%
Breakdown Viscosity (cP)	6.7	1036.7	198.3	−1781.3	−350	−210	−1280	99.6%
Final Viscosity (cP)	324	3969	3854	−2883	−2805	3370	−7483	98.1%
Setback Viscosity (cP)	56	2158	2312	−1755	−1991	3189	−6230	97.7%
Flow behavior index (n)	0.85	0.31	0.44	1.31	1.27	0.41	1.12	95.8%
Yield Stress (K)	5.72	7.19	6.83	−2.92	3.65	166.90	−367	97.4%

($p > 0.05$) of the independent variables of the cubic model adjusted for cupcake characteristics (TF = Teosinte Flour, BRF = High-Protein Brown Rice, WRF = High-Protein White Rice Flour, TF-BRF = Teosinte Flour with High-Protein Brown Rice, TF-WRF = Teosinte Flour with High-Protein White Rice Flour, BRF-WRF = High-Protein White Rice Flour with High-Protein Brown Rice Flour, TF-BRF-WRF = Teosinte Flour with High-Protein White Rice Flour and with High-Protein Brown Rice Flour).

4. Conclusions

This research examined the development of gluten-free bread by using TF, BRF, and WRF to observe the effects of the different sources on bread properties. Lower pasting properties, such as breakdown and setback viscosity, indicated that TF samples would be more stable. In bread, TF had negative effects on specific volume, texture, and color of breads, resulting in a greater hardness and greater crumb darkness, but these issues can be mitigated by using a combination of brown rice flour or white rice flour with teosinte flour to make the breads.

Author Contributions: R.S.A.: Conceptualization, data analysis and interpretation, writing; F.D.R.: conceptualization, performed most of the research work, M.S.M.; A.M.: data collection; S.K.P.: writing; Y.M.: formulated controls; W.P.: development and design of the experiments; J.M.K.: development and design of the experiments, supervision, project administration, writing and resource/funding acquisition, J.M.F.: supervision, project administration and resource/funding acquisition. All authors have read and agreed to the published version of the manuscript.

Funding: This work was supported by the Hatch fund at Universidad National de Agriculture (UNAG) with registration # AL-UNAG-028-2020. This work was also supported by the USDA National Institute of Food and Agriculture, Hatch project LAB94466.

Data Availability Statement: Data is contained within the article.

Conflicts of Interest: The authors declare no conflict of interest.

References

1. Cappelli, A.; Cini, E.; Guerrini, L.; Masella, P.; Angeloni, G.; Parenti, A. Predictive models of the rheological properties and optimal water content in doughs: An application to ancient grain flours with different degrees of refining. *J. Cereal Sci.* **2018**, *83*, 229–235. [CrossRef]
2. Ren, Y.; Linter, B.R.; Linforth, R.; Foster, T.J. A comprehensive investigation of gluten free bread dough rheology, proving and baking performance and bread qualities by response surface design and principal component analysis. *Food Funct.* **2020**, *11*, 5333–5345. [CrossRef] [PubMed]
3. Rej, A.; Aziz, I.; Sanders, D.S. Breaking bread! *Proc. Nutr. Soc.* **2019**, *78*, 118–125. [CrossRef] [PubMed]
4. Lebwohl, B.; Sanders, D.S.; Green, P.H.R. Coeliac disease. *Lancet* **2018**, *391*, 70–81. [CrossRef] [PubMed]
5. Bender, D.; Schönlechner, R. Innovative approaches towards improved gluten-free bread properties. *J. Cereal Sci.* **2020**, *91*, 102904. [CrossRef]
6. Capriles, V.D.; Arêas, J.A.G. Novel Approaches in Gluten-Free Breadmaking: Interface between Food Science, Nutrition, and Health. *Compr. Rev. Food Sci. Food Saf.* **2014**, *13*, 871–890. [CrossRef]
7. de la Hera, E.; Martinez, M.; Gómez, M. Influence of flour particle size on quality of gluten-free rice bread. *LWT—Food Sci. Technol.* **2013**, *54*, 199–206. [CrossRef]

8. Skendi, A.; Mouselemidou, P.; Papageorgiou, M.; Papastergiadis, E. Effect of acorn meal-water combinations on technological properties and fine structure of gluten-free bread. *Food Chem.* **2018**, *253*, 119–126. [CrossRef]
9. Roman, L.; Belorio, M.; Gomez, M. Gluten-Free Breads: The Gap Between Research and Commercial Reality. *Compr. Rev. Food Sci. Food Saf.* **2019**, *18*, 690–702. [CrossRef]
10. Wu, J.; McClements, D.J.; Chen, J.; Hu, X.; Liu, C. Improvement in nutritional attributes of rice using superheated steam processing. *J. Funct. Foods* **2016**, *24*, 338–350. [CrossRef]
11. Bastias-Montes, J.M.; Flores-Varela, L.E.; Reyes-Calderón, O.A.; Vidal-San-Martín, C.; Muñoz-Fariña, O.; Quevedo-León, R.; Acuña-Nelson, S.M. Teosinte (*Dioon mejiae*) flour: Nutritional and physicochemical characterization of the seed flour of the living fossil in honduras. *Agronomy* **2020**, *10*, 481. [CrossRef]
12. Comino, I.; de Lourdes Moreno, M.; Real, A.; Rodríguez-Herrera, A.; Barro, F.; Sousa, C. The Gluten-Free Diet: Testing Alternative Cereals Tolerated by Celiac Patients. *Nutrients* **2013**, *5*, 4250–4268. [CrossRef]
13. Hozyasz, K.K. Letter to the Editor Re: Comino, I.; et al. Nutrients 2013, 5, 4250–4268. *Nutrients* **2013**, *5*, 4964–4965. [CrossRef] [PubMed]
14. Saturni, L.; Ferretti, G.; Bacchetti, T. The Gluten-Free Diet: Safety and Nutritional Quality. *Nutrients* **2010**, *2*, 16–34. [CrossRef] [PubMed]
15. Amador, M.; Montilla, I.M.C.; Martín, C.S. Alternative grains as potential raw material for gluten—Free food development in the diet of celiac and gluten—Sensitive patients. *Austin J. Nutr. Metab.* **2014**, *2*, 1–9.
16. Bonta, M.; Pinot, O.; Graham, D.; Haynes, J.; Sandoval, G. Ethnobotany and conservation of tiusinte (*Dioon mejiae* Standl. & LO Williams, Zamiaceae) in northeastern Honduras. *J. Ethnobiol.* **2006**, *26*, 228–257.
17. Han, A.; Romero, H.M.; Nishijima, N.; Ichimura, T.; Handa, A.; Xu, C.; Zhang, Y. Effect of egg white solids on the rheological properties and bread making performance of gluten-free batter. *Food Hydrocoll.* **2019**, *87*, 287–296. [CrossRef]
18. Gujral, H.S.; Guardiola, I.; Carbonell, J.V.; Rosell, C.M. Effect of cyclodextrinase on dough rheology and bread quality from rice flour. *J. Agric. Food Chem.* **2003**, *51*, 3814–3818. [CrossRef] [PubMed]
19. Ribotta, P.D.; Ausar, S.F.; Morcillo, M.H.; Pérez, G.T.; Beltramo, D.M.; León, A.E. Production of gluten-free bread using soybean flour. *J. Sci. Food Agric.* **2004**, *84*, 1969–1974. [CrossRef]
20. Rostamian, M.; Milani, J.M.; Maleki, G. Physical properties of gluten-free bread made of corn and chickpea flour. *Int. J. Food Eng.* **2014**, *10*, 467–472. [CrossRef]
21. Smidova, Z.; Rysova, J. Gluten-free bread and bakery products technology. *Foods* **2022**, *11*, 480. [CrossRef] [PubMed]
22. Amdoun, R.; Khelifi, L.; Khelifi-Slaoui, M.; Amroune, S.; Asch, M.; Assaf-ducrocq, C.; Gontier, E. The Desirability Optimization Methodology; a Tool to Predict Two Antagonist Responses in Biotechnological Systems: Case of Biomass Growth and Hyoscyamine Content in Elicited Datura starmonium Hairy Roots. *Iran. J. Biotechnol.* **2018**, *16*, 11–19. [CrossRef]
23. AACC. *Approved Methods of the AACC: Method 61.02.01 (Pasting Properties)*, 11th ed.; American Association of Cereal Chemists: St. Paul, MN, USA, 2012.
24. Ye, J.; Wang, J.; Zhu, Y.; Wei, Q.; Wang, X.; Yang, J.; Tang, S.; Liu, H.; Fan, J.; Zhang, F.; et al. A thermoresponsive polydiolcitrate-gelatin scaffold and delivery system mediates effective bone formation from BMP9-transduced mesenchymal stem cells. *Biomed. Mater.* **2016**, *11*, 025021. [CrossRef] [PubMed]
25. AACC. *Approved Methods of the AACC*, 10th ed.; Methods of 10-05, 10-91, 44-15, and 74-09; American Association of Cereal Chemists: St. Paul, MN, USA, 2000.
26. Kadan, R.S.; Robinson, M.G.; Thibodeaux, D.P.; Pepperman, A.B. Texture and other physicochemical properties of whole rice bread. *J. Food Sci.* **2001**, *66*, 940–944. [CrossRef]
27. Aleman, R.S.; Paz, G.; Prinyawiwatkul, W.; Moncada, M.; King, J.M. Comparison of the Thermal and Rheological Properties of Frontière Brown Rice Flour, Tapioca Starch, and Potato Starch and Mixture Effects on Pasting Properties in Aqueous Systems. *Starch-Stärke* **2023**, 2200196. [CrossRef]
28. Aleman, R.S.; Paz, G.; Morris, A.; Prinyawiwatkul, W.; Moncada, M.; King, J.M. High protein brown rice flour, tapioca starch & potato starch in the development of gluten-free cupcakes. *LWT* **2021**, *152*, 112326. [CrossRef]
29. Monteiro, J.S.; Farage, P.; Zandonadi, R.P.; Botelho, R.B.A.; de Oliveira, L.D.L.; Raposo, A.; Shakeel, F.; Alshehri, S.; Mahdi, W.A.; Araújo, W.M.C. A systematic review on gluten-free bread formulations using specific volume as a quality indicator. *Foods* **2021**, *10*, 614. [CrossRef] [PubMed]
30. Ziobro, R.; Juszczak, L.; Witczak, M.; Korus, J. Non-gluten proteins as structure forming agents in gluten free bread. *J. Food Sci. Technol.* **2016**, *53*, 571–580. [CrossRef]
31. Paz, G.M.; King, J.M.; Prinyawiwatkul, W. High Protein Rice Flour in the Development of Gluten-Free Bread. *J. Culin. Sci. Technol.* **2021**, *19*, 315–330. [CrossRef]
32. Korus, A.; Witczak, M.; Korus, J.; Juszczak, L. Dough Rheological Properties and Characteristics of Wheat Bread with the Addition of Lyophilized Kale (*Brassica oleracea* L. var. *sabellica*). *Powder. Appl. Sci.* **2022**, *13*, 29. [CrossRef]
33. Sanjyal, S.; Hampton, J.G.; Rolston, P.; Marahatta, S. Teosinte (*Euchlaena mexicana* L.) Seed Production: Effect of Sowing Date, Seed Rate and Cutting Management on Seed Yield. *Agronomy* **2022**, *12*, 1646. [CrossRef]
34. Lamberts, L.; Delcour, J.A. Carotenoids in raw and parboiled brown and milled rice. *J. Agric. Food Chem.* **2008**, *56*, 11914–11919. [CrossRef] [PubMed]

35. Leon, E.; Piston, F.; Aouni, R.; Shewry, P.R.; Rosell, C.M.; Martin, A.; Barro, F. Pasting properties of transgenic lines of a commercial bread wheat expressing combinations of HMW glutenin subunit genes. *J. Cereal Sci.* **2010**, *51*, 344–349. [CrossRef]
36. Zhou, D.N.; Zhang, B.; Chen, B.; Chen, H.Q. Effects of oligosaccharides on pasting, thermal and rheological properties of sweet potato starch. *Food Chem.* **2017**, *230*, 516–523. [CrossRef]
37. Aleman, R.S.; Morris, A.; Prinyawiwatkul, W.; Moncada, M.; King, J.M. Physicochemical properties of Frontière rice flour and its application in a gluten-free cupcake. *Cereal Chem.* **2022**, *99*, 303–315. [CrossRef]
38. Singh Sandhu, K.; Singh, N.; Lim, S.T. A comparison of native and acid thinned normal and waxy corn starches: Physicochemical, thermal, morphological and pasting properties. *LWT* **2007**, *40*, 1527–1536. [CrossRef]
39. Ragaee, S.; Abdel-Aal, E.S.M. Pasting properties of starch and protein in selected cereals and quality of their food products. *Food Chem.* **2006**, *95*, 9–18. [CrossRef]
40. Ohizua, E.R.; Adeola, A.A.; Idowu, M.A.; Sobukola, O.P.; Afolabi, T.A.; Ishola, R.O.; Ayansina, S.O.; Oyekale, T.O.; Falomo, A. Nutrient composition, functional, and pasting properties of unripe cooking banana, pigeon pea, and sweetpotato flour blends. *Food Sci. Nutr.* **2017**, *5*, 750–762. [CrossRef]

foods MDPI

Article

Developing a Clean Labelled Snack Bar Rich in Protein and Fibre with Dry-Fractionated Defatted Durum Wheat Cake

Giacomo Squeo, Vittoria Latrofa, Francesca Vurro, Davide De Angelis, Francesco Caponio, Carmine Summo and Antonella Pasqualone *

Department of Soil, Plant and Food Science (DISSPA), University of Bari Aldo Moro, Via Amendola, 165/a, I-70126 Bari, Italy; giacomo.squeo@uniba.it (G.S.); vittoria.latrofa@uniba.it (V.L.); francesca.vurro@uniba.it (F.V.); davide.deangelis@uniba.it (D.D.A.); francesco.caponio@uniba.it (F.C.); carmine.summo@uniba.it (C.S.)
* Correspondence: antonella.pasqualone@uniba.it

Abstract: The shift towards a vegetarian, vegan, or flexitarian diet has increased the demand for vegetable protein and plant-based foods. The defatted cake generated during the extraction of lipids from durum wheat (*Triticum turgidum* L. var. *durum*) milling by-products is a protein and fibre-containing waste, which could be upcycled as a food ingredient. This study aimed to exploit the dry-fractionated fine fraction of defatted durum wheat cake (DFFF) to formulate a vegan, clean labelled, cereal-based snack bar. The design of experiments (DoEs) for mixtures was applied to formulate a final product with optimal textural and sensorial properties, which contained 10% DFFF, 30% glucose syrup, and a 60% mix of puffed/rolled cereals. The DFFF-enriched snack bar was harder compared to the control without DFFF (cutting stress = 1.2 and 0.52 N/mm^2, and fracture stress = 12.9 and 9.8 N/mm^2 in the DFFF-enriched and control snack bar, respectively), due to a densifying effect of DFFF, and showed a more intense yellow hue due to the yellow–brownish colour of DFFF. Another difference was in the caramel flavour, which was more intense in the DFFF-enriched snack bar. The nutritional claims "low fat" and "source of fibre" were applicable to the DFFF-enriched snack bar according to EC Reg. 1924/06.

Keywords: cereal based; texture; upcycling; ready to eat food; design of experiments; texture analysis; nutritional composition

Citation: Squeo, G.; Latrofa, V.; Vurro, F.; De Angelis, D.; Caponio, F.; Summo, C.; Pasqualone, A. Developing a Clean Labelled Snack Bar Rich in Protein and Fibre with Dry-Fractionated Defatted Durum Wheat Cake. *Foods* **2023**, *12*, 2547. https://doi.org/10.3390/foods12132547

Academic Editors: Jennifer Ahn-Jarvis and Brittany A. Hazard

Received: 15 June 2023
Revised: 27 June 2023
Accepted: 28 June 2023
Published: 29 June 2023

1. Introduction

One third of food produced for human consumption is lost or wasted worldwide, accounting for about 1.3 billion tons annually [1]. Alongside this, food supply chains generate a large amount of valuable by-products that still could be exploited and valorised for human nutrition. As for the supply chain of durum wheat (*Triticum turgidum* L. var. *durum*), during the milling process, a by-product is generated composed of germ, bran, and debranning fractions [2]. Being rich in high biological value protein, minerals, fibre, and lipids, this by-product can be further exploited [3,4]. Indeed, the extraction of the lipid fraction, rich in vitamins and essential fatty acids, has been proposed [5], and durum wheat oil has been used for the preparation of biscuits and focaccia [6,7]. The lipid extraction, however, generates a second by-product, namely, the defatted durum wheat cake. To upcycle the defatted cake, after stripping and micronisation, the dry-fractionation process can be applied, allowing a coarse fraction rich in starch and fibre and a fine fraction rich in protein to be obtained [8]. The coarse fraction of dry-fractionated cereal flours has already been proposed as an ingredient in different food products, such as meat analogues and spaghetti [9,10]. In addition, the fine fraction can also be used to fortify foods [11] with the advantage of being derived by a more sustainable method of protein enrichment compared to a wet concentration, which uses an alkaline extraction followed by isoelectric precipitation [8].

The sustainability of food systems is a hot topic increasingly considered by researchers, national/international organizations, and consumers. Additionally, the recovery of protein from vegetable waste and by-products matches the general request to replace animal protein with alternative plant protein [12]. In fact, the shift towards a vegetarian, vegan, or flexitarian diet has increased the demand for vegetable protein and plant-based foods [13,14]. The reduction of animal protein is associated with well-being and health, both being drivers that play a pivotal role in influencing the food choices of many consumers who integrate nutrition with sport activities [15]. To support this healthy lifestyle, consumers increasingly choose foods rich in protein and fibre, usually marketed in the form of snack bars [16,17]. In addition, the fast pace of life further concurs to the "snacking" effect, mostly based on the consumption of snack bars.

Given that sweet snacks are considered unhealthy due to the presence of high levels of sugars, food companies are releasing snacks formulated in a healthier way, incorporating protein isolates to increase protein content [18] and preferring clean labelled formulations [19,20]. Though specific regulations/legislations of clean labels do not exist, and a clear definition has not been established [21], this new trend in food products, mostly referring to the absence of additives, has been taken up by a multitude of food industry stakeholders [22]. Indeed, many additives can be included in the formulation of snack bars. Among the most used, sorbitol and polyols, responsible for a laxative effect, are sugar replacers, while lecithin is the most common emulsifier, and glycerol is added as a gelatinizing agent, followed by colouring agents and artificial flavours to improve the sensory properties [23]. Moreover, saturated fats, involved in cholesterol increase and cardiovascular disease, are reported among the variety of ingredients commonly used in snack bars [23].

Over the past few years, several studies have been conducted concerning the re-use of food industry by-products (such as the bran of black rice and corn, banana peel powder, sunflower meal, and jackfruit seed flour) in the formulation of snack bars [20,24–27]. These studies have confirmed the increasing interest towards this food category, highlighting its suitability as a recipient for fortification and functionalization. However, the re-use of defatted durum wheat cake, which can be easily dry-fractionated to obtain a protein-rich flour, has not been considered so far.

In this framework, this study aimed at exploiting the dry-fractionated fine fraction of defatted durum wheat cake (DFFF) to formulate a vegan, clean labelled, cereal-based snack bar, evaluating the influence of the ingredients on the textural, sensory, and nutritional properties of the final product.

2. Materials and Methods

2.1. Basic Ingredients

DFFF was provided by Casillo Next Gen Food srl (Corato, Italy). Its preparation was as follows: The durum wheat milling by-product was composed of a mix of germ, bran, and debranning fractions, and it was submitted to the extraction of the oily fraction using *n*-hexane, as described in the study by Squeo et al. (2022) [5]. The remaining cake was micronized using an impact mill (UPZ 100, Hosokawa-Alpine, Augusta, Germany). The mill speed was set at 15,000 rpm and the feed rate at 3 kg h^{-1}. Then, an air classifier (ATP 50, Hosokawa-Alpine, Augusta, Germany) was used to dry-fractionate the micronized cake flour into coarse and fine fractions. The speed of the grader wheel was set at 3000 rpm and the air flow rate at 50 m^3 h^{-1}. The fine fraction was recovered, having an *n*-hexane residual content <5 ppm, i.e., food grade. The chemical composition of DFFF is shown in Figure 1.

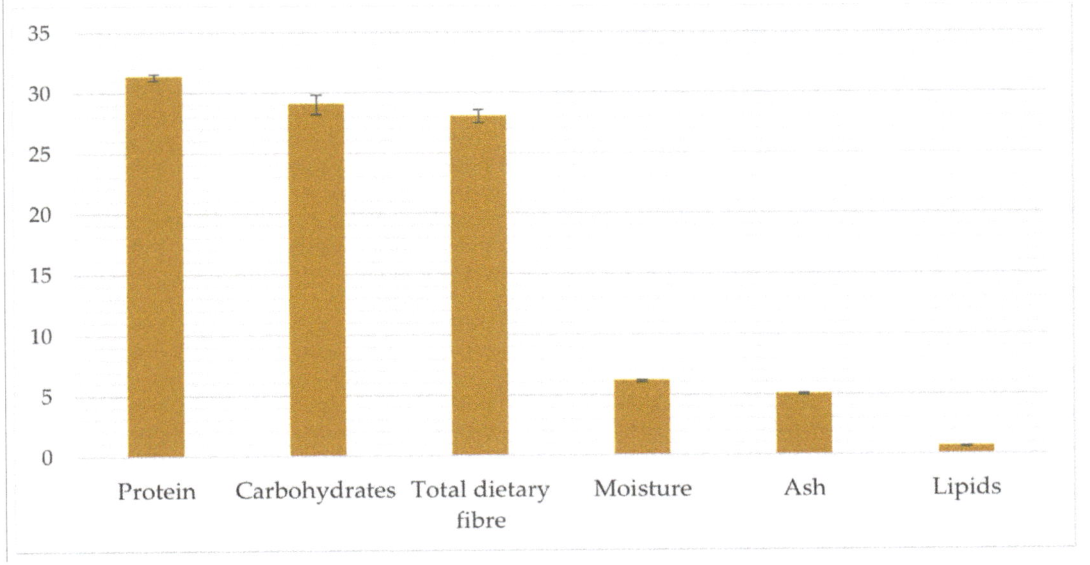

Figure 1. Proximate composition (g 100 g^{-1}) of dry-fractionated fine fraction of defatted durum wheat cake (DFFF).

Puffed hulled millet (Vivibio, Villareggia, Italy), 5-grain rolled cereals (20% oat, 20% barley, 20% rice, 20% rye, 20% wheat) (Fiorentini, Troffarello, Italy), and puffed spelt (Selex, Trezzano, Italy) were purchased from a local retailer.

2.2. Experimental Design and Preparation of the Snack Bars

The design of experiments (DoEs) for mixtures [28] was applied to formulate a product with optimal textural and sensorial properties. Three factors were considered, which were the three main ingredients, namely, DFFF, glucose syrup, and a cereal mix composed of puffed hulled millet, 5-grain rolled cereals, and puffed spelt (35:40:25 *w*/*w*/*w*). The quantitative ranges were 10–50%, 30–60%, and 20–60% for DFFF, glucose syrup, and cereal mix, respectively. Then, from all the possible mixtures in the defined experimental domain, 8 unique formulations were chosen (Table 1) by a D-optimality criterion [29] to model a special cubic response.

Table 1. List of the experiments performed.

Experiment	Dry-Fractionated Fine Fraction of Defatted Durum Wheat Cake—X1 (g 100 g^{-1})	Glucose Syrup—X2 (g 100 g^{-1})	Puffed/Rolled Cereal Mix—X3 (g 100 g^{-1})
1	50	30	20
2	20	60	20
3	10	60	30
4	10	30	60
5	10	45	45
6	35	45	20
7	25	40	35
8	30	30	40

To prepare the snack bars, glucose syrup (Ambrosio, Striano, Italy) and DFFF were manually mixed and heated in a pan on an induction heater (G3 Ferrari, Rimini, Italy) at the power of 2000 W until the complete formation of a cream was reached (about 5 min). Then, the cereal mix was added and manually mixed to distribute it homogeneously into the cream, followed by pouring into a silicone mould (FantasyDay, Changan Town, China) having six 2 cm thick cavities of 11.5 × 3.8 cm, and cooking in a ventilated oven (Smeg, Guastella, Italia) for 5 min at 120 °C. The snack bars were allowed to cool down at room temperature (20 ± 1 °C) and then extracted from the moulds.

Three snack bars for each formulation of the experimental design were prepared (n = 3). A control snack bar, without DFFF, was prepared with 33.3 g 100 g^{-1} of glucose syrup and 66.6 g 100 g^{-1} of cereal mix.

2.3. Determination of the Chemical Composition

Protein content (N × 5.7) was determined according to the method AOAC 979.09 [30]. The lipid content was determined using a Soxhlet apparatus, according to the method AOAC 945.38F [30], using diethyl ether as the extraction solvent. The total dietary fibre was analysed by the enzymatic–gravimetric procedure as described in the method AOAC 985.29 [30]. Moisture and ash content were determined according to the AOAC methods 925.10 and 923.03 [30], respectively. The available carbohydrate content was determined by difference, subtracting the content of protein, lipid, fibre, moisture, and ash to 100. The energy value was calculated by using Atwater general conversion factors. All determinations were carried out in triplicate.

2.4. Texture Analysis

The textural properties of the snack bars were determined by a cutting test and a three-point bending test. A texture analyser, equipped with a 1 kN load cell (Z1.0 TN, Zwick Roell, Ulm, Germany), was used for both tests.

The cutting test was carried out as described in Costantini et al. [31] with some modification. The sample (11.5 × 3.8 × 2 cm, L × W × H) was positioned on the base in order to be cut by the 50 mm length blade on its short side positioned 25 mm from the surface of the sample. The blade descended at a speed of 1 mm/s until the sample was cut. The cutting stress, i.e., the maximum force applied on the surface until the sample was completely cut (N/mm^2), and the deformation until the cut (mm) were measured.

Three-point bending was carried as reported in Pasqualone et al. [32] with a few modifications. The texture analyser was equipped with a 1 kN load cell, a metal breaking probe, and two metal supports, 6 cm apart from each other, to hold the whole sample (11.5 × 3.8 × 2 cm, L × W × H). The sample was placed on the two supports, and the probe (positioned 25 mm above the sample) descended on it at a speed of 5 mm/s. The fracture stress (N/mm^2), i.e., the maximum force required to break the sample, and the deformation until rupture (mm) were measured.

Data acquisition was accomplished using TestXPertII v3.41 software (Zwick Roell, Ulm, Germany). The analysis was carried out in triplicate.

2.5. Sensory Analysis

The sensory features were evaluated by a panel consisting of 8 trained members aged between 26 and 56 years. The panellists did not suffer from any food intolerances or allergies, received information on the objectives of the study, and provided written informed consent.

At first, the snack bars corresponding to the 8 different formulations (Table 1) were evaluated on a 0–9 score range (0 = minimum; 9 = maximum) for their stickiness (a critical parameter for these products) and acceptability. The results of this evaluation, together with the textural data, allowed the optimal snack bar to be selected. Then, both the optimal and the control snack bar were submitted to Quantitative Descriptive Analysis (QDA) for a detailed sensory profiling. Eight sensory descriptors were evaluated on a 0–9 score range

(0 = minimum; 9 = maximum intensity) regarding visual appearance (homogeneity of the distribution of cereals, brightness, yellowish colour), flavour (cereal-like, toasted, caramel and abnormal), and taste (sweet).

2.6. Colour Analysis

The snack bars had a highly heterogeneous colour that could not be analysed directly with a colourimeter, and, for this reason, image analysis and colour elaboration was carried out. An RGB image was collected using a Canon EOS 600D camera equipped a Canon EF-S 18–200mm f/3.5–5.6 IS lens (Canon Inc., Tokyo, Japan) at a 50 cm distance between the lens and the sample. The optimal and control snack bars, placed inside a white plastic box, sized 49 × 51 × 50 cm, with a black base, were photographed from above. The image (2193 × 2683 pixels) was imported into MATLAB R2021a (The MathWorks Inc., Natick, MA, USA) and processed as follows: After masking and background removal, carried out using the HYPER-Tools toolbox (freely available at www.hypertools.org, accessed on 20 April 2023) [33], the image was converted from RGB colour space to $L^*a^*b^*$ colour space, thus obtaining the respective values of brightness and red and yellow indices per pixel. Then, the absolute frequency distribution and the cumulative frequency distribution of the pixels were calculated per each $L^*a^*b^*$ index to highlight differences in the colour characteristics of the snack bars.

2.7. Statistical Analysis

The mixture experimental design was set up and analysed using CAT (Chemometric Agile Tool) software [34], freely downloadable from http://gruppochemiometria.it/index.php/software, accessed on 20 April 2023. All data were expressed as the mean ± standard deviation (n = 3). One-way analysis of variance (ANOVA) was performed, followed by the Tukey test for multiple comparisons using Minitab 17 (Minitab Inc., State College, PA, USA).

3. Results and Discussion

3.1. Identification of the Optimal Formulation

Figure 2 depicts the contour plots of the cutting stress (A), the fracture stress (B), the sensory acceptability (C), and the stickiness (D) of the experimental snack bars in the defined domain. DoEs allowed us to reach a comprehensive understanding of the ingredients' effects on the snack bar properties and to select the better formulation. A snack bar, indeed, is a mixture of ingredients that substantially retain their original physical characteristics, which are generally very different from each other, so their relative quantities greatly influence the features of the final product.

The cutting and fracture stress, determined by measuring the maximum force required to cut or break the sample, respectively (Figure 2A,B), were considered because they are known to be related to sensory firmness and stickiness to the teeth [35], which are important characteristics of snack bars. The fitting of the responses, however, was not excellent, showing in both the cases an R^2 value equal to 0.6. This was likely due to the high physical inhomogeneity of this kind of product, which brought great variability in the responses, making the modelling more difficult. Nonetheless, from the contour plots, useful information about the effects of the ingredients on the mechanical features of the snack bar could be caught. Indeed, the cutting and fracture stress values, and thus the hardness of the snack bars, progressively increased by raising the proportion of X2 (glucose syrup). Therefore, the highest amount of glucose syrup gave rise to very hard bars, as observed close to the X2 vertex. On the other hand, the experimental region close to the X1 vertex corresponded to the worst results due to a total lack of consistency. The snack bars in this region (trials 1 and 8) were impossible to shape properly, showing a powdery and poor structure (Figure 3). Experiments 1 (50% DFFF, 30% glucose syrup, 20% cereals) and 8 (30% DFFF, 30% glucose syrup, 40% cereals) highlighted that the snack bar could not be formed when DFFF was ≥ glucose syrup. Indeed, DFFF was rich in protein and fibre, so its increase would be nutritionally valuable [36], but being a powdery ingredient, an excessive amount disrupted the structure of the snack bar, especially when the amount of glucose syrup was too low. At the same time, high amounts of glucose syrup excessively increased the hardness. By increasing the amount of the cereal mix (X3), a moderate and positive effect on the structure was observed. Therefore, for the textural features, the best experiments were those in the centre of the domain (point 7), at the X3 vertex (point 4), and along the edge of X3–X2 (points 3 and 5) (Figure 2A,B).

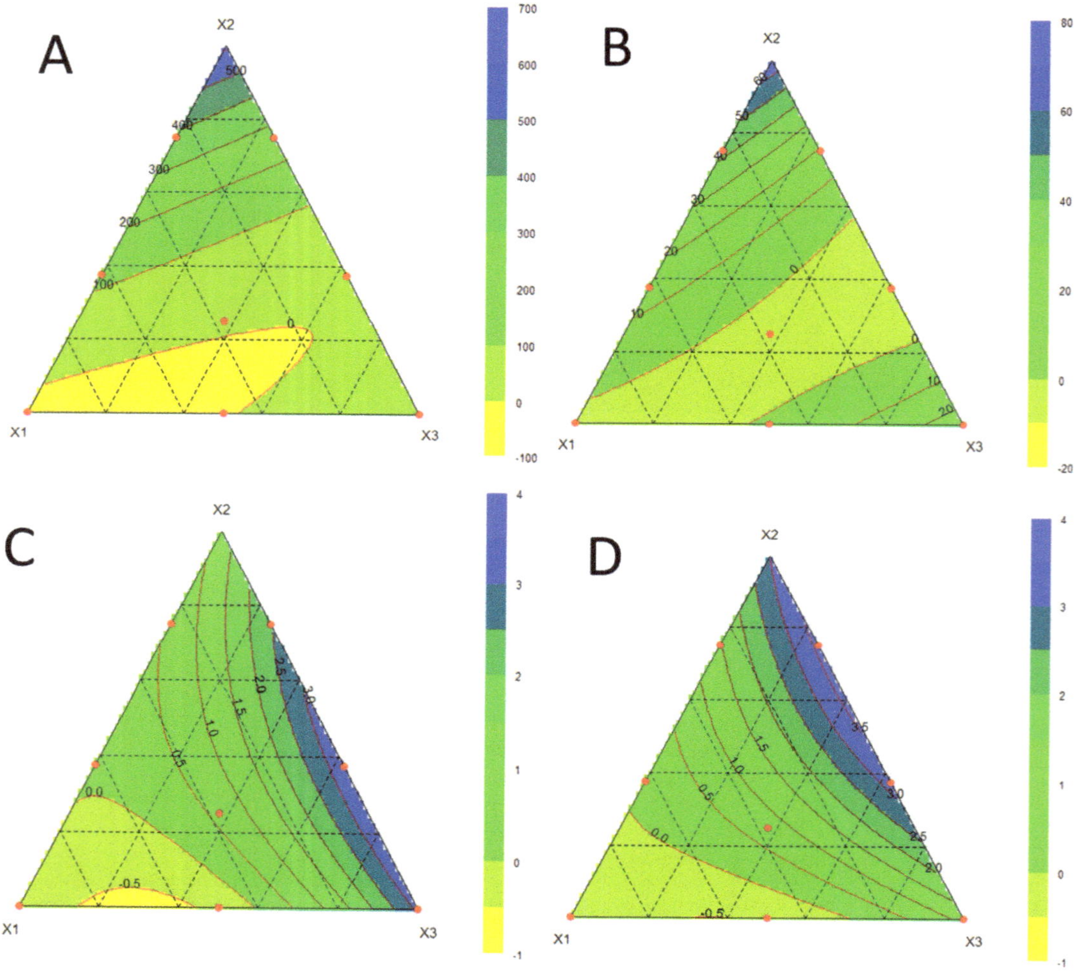

Figure 2. Contour plots of the cutting stress (N/mm^2, (**A**)), fracture stress (N/mm^2, (**B**)), sensory acceptability (**C**), and stickiness (**D**) of the experimental snack bars. X1 = dry-fractionated fine fraction of defatted durum wheat cake (DFFF); X2 = glucose syrup; X3 = puffed/rolled cereal mix. Red point represents the experiments carried out.

Figure 3. Experimental snack bars formulated with dry-fractionated fine fraction of defatted durum wheat cake (DFFF), glucose syrup, and puffed/rolled cereal mix. Numbers 1–8 correspond to the formulations reported in Table 1.

Together with the textural properties, two other parameters were chosen for the selection of the optimal formulation, i.e., sensory acceptability, which is fundamental for food products in general, and stickiness, which is important especially for the snack bars. The responses for these two parameters are depicted in Figure 2C,D, respectively. In this case, the fitting of the responses was better than the textural parameters, showing an R^2 value equal to 0.9 for the sensory acceptability and 0.7 for the stickiness, both markedly influenced by variations in the ingredient ratio. The stickiness was very high in the upper part of the X3–X2 edge (experiments 3 and 5). The highest sensory acceptability and the lowest stickiness were observed in the experimental region close to the X3 vertex (point 4).

In light of these results, the most suitable trade-off was considered to be the formulation corresponding to the X3 vertex (experiment 4), i.e., 10% DFFF, 30% glucose syrup, and 60% cereals. This well-structured bar, neither too sticky nor too hard, was the optimal one and was then further characterized for its nutritional, textural, and sensorial features.

3.2. Nutritional, Textural, and Sensorial Characterization

Table 2 shows the results of the nutritional composition of the DFFF-enriched snack bar, compared with the control snack bar, prepared without DFFF.

Table 2. Nutritional composition of the experimental snack bars. DFFF = dry-fractionated fine fraction of defatted durum wheat cake.

Parameter	DFFF-Enriched Snack Bar			Control Snack Bar		
	Amount per 100 g	Amount per Unit (33 g)	% Daily Value per Unit (33 g) *	Amount per 100 g	Amount per Unit (33 g)	% Daily Value per Unit (33 g) *
Moisture (g)	8.3 ± 0.11 [b]	2.7	-	9.3 ± 0.17 [a]	3.2	-
Protein (g)	9.7 ± 0.24 [a]	3.2	6.4	6.5 ± 0.08 [b]	2.3	4.3
Total dietary fibre (g)	3.2 ± 0.20 [a]	1.1	4.2	2.3 ± 0.31 [b]	0.8	3.0
Lipids (g)	1.4 ± 0.09 [a]	0.5	0.7	1.4 ± 0.04 [a]	0.5	0.7
Carbohydrates (g)	76.4 ± 0.36 [b]	25.2	9.8	80.1 ± 0.63 [a]	26.4	10.2
Energy value (kcal)	363	120	6.0	364	120	6.0

* Energy, protein, lipids, and carbohydrates were calculated according to the daily reference intakes reported in the Annex XIII of the EU Reg. 1169/2011 [37]; total dietary fibre was calculated according to the daily reference intake reported in the EFSA recommendation [38]. Different letters indicate significant differences at $p < 0.05$.

The addition of DFFF caused a significant reduction of the moisture and an increase in protein content with respect to the control. These results were expected, DFFF being an ingredient high in protein and low in moisture. In particular, the increase in protein highlighted the effectiveness of using a dry-fractionated ingredient considering that studies involving the addition of other cereal by-products, such as black rice bran and maize bran [20,26,39], did not achieve a significant increase in protein content. One DFFF-enriched snack bar (weight = 33 g) covered 6.4% of the daily reference intake of protein, while the control snack bar provided only 4.3% of the reference intake. The consumption of plant-based protein has a positive effect on health, since it minimizes cardiometabolic risk factors and is inversely correlated with hypertension, obesity, and insulin resistance [40].

DFFF is also rich in fibre (Figure 1). Therefore, another positive result of the incorporation of DFFF was the enrichment in total dietary fibre (Table 2), recommended to improve the metabolism of gut microbiota and to reduce the risks connected with diabetes, obesity, and dyslipidaemia [41]. Moreover, the reached amount of fibre makes possible the labelling of the DFFF-enriched snack bar as a "source of fibre" according to the current rules [42], as it meets the minimum requirement of 3 g of fibre per 100 g of product. This amount was higher than the level reached by adding other by-products of the food industry, such as pumpkin seed flour or carrot powder [43,44]. Specifically, a single DFFF-enriched snack bar (33 g) provided 4.2% of the daily intake recommended by the EFSA for dietary fibre (i.e., 25 g) [38], while the control bar provided 3.0% of the reference intake.

The lipid content did not show a significant difference between the two snack bars. The observed amount makes possible the labelling of them as "low fat" according to the EC Reg. 1924/2006 [42] (no more than 3 g of fat in 100 g). The observed lipid content was considerably lower than in snack bars formulated with other by-products, such as tilapia frames or roasted rice bran [39,45].

Carbohydrates remained the main constituents of the snack bars, but they were significantly lower in the DFFF-enriched bar than in the control.

The two types of snack bars showed very similar energy values, providing 6% of the daily reference value of 2000 kcal [37].

As for the texture analysis of the snack bars, both cutting and bending tests, which helped also in selecting the optimal formulation, highlighted significant textural differences between the DFFF-enriched snack bar and the control, with the former being harder than the latter. The results are shown in Table 3.

Table 3. Texture analysis of the experimental snack bars. DFFF = dry-fractionated fine fraction of defatted durum wheat cake.

Parameter	Type of Snack Bar	
	DFFF-Enriched	**Control**
Cutting test		
Cutting stress (N/mm^2)	1.2 ± 0.1 ^{a}	0.52 ± 0.2 ^{b}
Deformation (mm)	16.3 ± 0.29 ^{a}	14.7 ± 0.56 ^{b}
Three point-bending test		
Fracture stress (N/mm^2)	12.9 ± 3.9 ^{a}	9.8 ± 4.09 ^{b}
Deformation (mm)	3.6 ± 0.8 ^{a}	2.5 ± 1.07 ^{b}

Different letters indicate significant differences at $p < 0.05$.

The observed increase in cutting and fracture stress was likely due to the densifying effect of DFFF when added to glucose syrup, leading to a thick cream that made the snack bar become very compact and hard. These results were correlated with the higher extent of deformation observed in the DFFF-enriched bar with respect to the control. This indicated a greater effort needed to chew the DFFF-enriched snack bar compared to the control and, more in general, evidenced the ability of the new formulation to tolerate higher stress. Foods that need longer chewing could reduce the appetite and then also the energy intake [46]. Therefore, a harder texture is desirable from a nutritional point of view. Zulaikha et al. [45] and Silva et al. [47] observed a similar increase in hardness following the addition of powdery ingredients, such as tilapia frames and marolo pulp flour, to snack bars.

As for the colour of the snack bars, high heterogeneity was found along their surfaces, as expected in this kind of product (Figure 3). As a consequence, a direct colorimeter evaluation was not sufficiently reliable, justifying the need for imaging approaches to better evaluate the product's colour. Figure 4A shows the image of the snack bars reconstructed according to the $L^*a^*b^*$ values of each pixel. In both the snack bars, the surface of the cereal grains showed higher values of brightness (L^*), which could be explained by the shiny coating formed by the glucose syrup. The pixel distribution of the L^* value (Figure 4B) was bimodal, i.e., presented two main peaks, indicating a strong contrast within the surface brightness in different parts of the snack bar due to the heterogeneity of the assembled ingredients. The curves of the cumulative distribution showed that the control bar had a higher percentage of pixels with high L^* values with respect to the DFFF-enriched snack bar, supporting the empirical observation (Figure 4A) that the former was, overall, brighter than the latter.

High rates of negative a^* values were recorded in both snack bars (Figure 4A), indicating a general trend toward the greenish colour, with some surface portions characterized by positive values and thus reddish features. The pixel distribution (Figure 4B) was very similar between the two snack bars, showing minor differences in the whole red–green feature, but with a higher proportion of reddish pixels in the DFFF-enriched snack bar with respect to the control.

A remarkable difference was observed for the b^* index, with the DFFF-enriched snack bar clearly presenting a more intense yellow hue (i.e., positive values) with respect to the control bar (Figure 4A). The DFFF ingredient, indeed, was characterized by a yellow–brownish colour that was reflected in a more yellow bar. This empirical observation was supported by the frequency distribution (Figure 4B) that clearly showed a higher proportion of pixels having high b^* values in the DFFF-enriched snack bar with respect to the control.

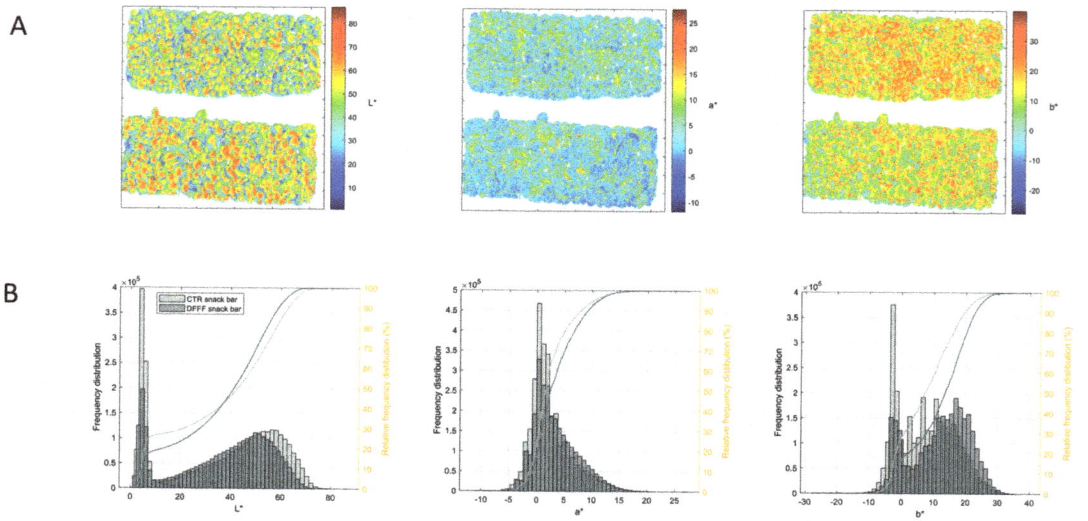

Figure 4. (**A**) L^*, a^*, and b^* reconstructed images of the DFFF-enriched snack bar (snack bar on the top of each image) and the control snack bar (snack bar on the bottom of each image). (**B**) Pixel absolute frequency distribution (bars) and cumulative frequency distribution (lines) of the control snack bar (light grey) and of the DFFF-enriched snack bar (dark grey) for L^*, a^*, and b^*.

Overall, it could be concluded that the DFFF-enriched snack bar was slightly less bright but redder and yellower than the control bar.

The QDA results are reported in Table 4. The visual evaluation revealed statistically significant differences in the colour, with the DFFF-enriched snack bar perceived as more yellow than the control, in agreement with the instrumental analysis. Brightness was also in line with the instrumental results because the DFFF-enriched bar was significantly less bright than the control, probably because the addition of DFFF diminished the glossy appearance of the surface.

Table 4. Results of the quantitative descriptive sensory analysis of the experimental snack bars. DFFF = dry-fractionated fine fraction of defatted durum wheat cake.

Sensory Parameter	Type of Snack Bar	
	DFFF-Enriched	**Control**
Visual		
Yellowish colour	6.1 ± 1.2^a	4.0 ± 0.9^b
Brightness	5.0 ± 1.1^b	7.1 ± 0.8^a
Homogeneity of the distribution of cereals	6.8 ± 0.4^a	7.0 ± 0.6^a
Flavour		
Cereal-like	6.8 ± 0.9^a	6.8 ± 1.3^a
Toasted	4.7 ± 1.2^a	3.8 ± 1.0^a
Caramel	4.1 ± 0.7^a	2.2 ± 1.1^b
Abnormal odours	0.2 ± 0.4^a	0.3 ± 0.8^a
Taste		
Sweetness	4.5 ± 0.5^a	4.1 ± 0.5^a

Different letters indicate significant differences at $p < 0.05$.

From a practical point of view, the corroboration of the results of the instrumental colour evaluation and of the sensory evaluation suggests that image processing technology may be implemented on the production process for quality control procedures, considering also that colour determination with classical colorimeters is challenging for such heterogeneous products, and the sensory evaluation requires trained panellists and is time consuming. On the contrary, no significant differences emerged in the homogeneity of the distribution of cereals.

Flavour attributes were ascribed to the typical characteristic of cereals and/or connected to the cooking step. The addition of the DFFF did not affect the cereal-like note. The flavour of caramel, instead, was significantly more intense in the DFFF-enriched snack bar than in the control, probably due to the thermal effect exerted by the hot glucose syrup on DFFF during the preparation of the snack bar. Heat treatments of bran and similar materials are known to develop volatile compounds with sweet and caramel-like aromas [48]. Indeed, also the toasted note was perceived more intensely in the DFFF-enriched snack bar, but the difference with the control was not statistically significant. No abnormal smells were detected in either samples. The taste evaluation showed that both the snack bars had an equally perceived medium sweet taste, imputable to the glucose syrup.

4. Conclusions

The use of DFFF as a food ingredient could favour the circularity of the durum wheat supply chain while responding to the demand for healthy and ready-to-eat food products expressed by modern consumers. The outcomes of this study demonstrated that the development of an innovative snack bar enriched with DFFF is a feasible strategy to this end.

DFFF should not be added in quantities higher than 10%, and in any case must be lower than the amount of glucose syrup in order to ensure an optimal structure and adequate sensory acceptability of the snack bar. Nonetheless, thanks to the fortification with DFFF, the proposed snack bar showed improved nutritional characteristics, with enhanced protein content. Noteworthily, the optimized formulation can be labelled as "low fat" and a "source of fibre" according to the EC Reg. 1924/06. Furthermore, being exclusively made of plant-based ingredients, without additives, the DFFF-enriched snack bar was a clean-labelled food product and met the expectations of vegetarian and vegan consumers.

This study could pave the way for the formulation of other DFFF-enriched snack bars, for example, by incorporating nuts and/or dehydrated fruit, to offer the consumer a wide choice of flavours and tastes while meeting the current trends in sustainable food consumption.

Author Contributions: Conceptualization, A.P.; methodology, V.L. and F.V.; software, G.S.; formal analysis, G.S., D.D.A. and C.S.; investigation, V.L. and F.V.; data curation, G.S. and V.L.; writing—original draft preparation, G.S. and V.L.; writing—review and editing, A.P., F.C. and C.S.; visualization, V.L.; supervision, A.P. All authors have read and agreed to the published version of the manuscript.

Funding: This research received no external funding.

Data Availability Statement: The data used to support the findings of this study can be made available by the corresponding author upon request.

Acknowledgments: The authors would like to thank José Manuel Amigo of Bilbao IKERBASQUE Basque Foundation for Science (Bilbao, Spain) and the Department of Analytical Chemistry, University of the Basque Country (Bilbao, Spain) for help with the image processing, and Casillo Next Gen Food s.r.l. (Corato, Italy) for kindly providing the dry-fractionated fine fraction of defatted durum wheat cake. The authors also thank Marcello Greco Miani (Casillo Next Gen Food s.r.l., Corato, Italy) for valuable discussions.

Conflicts of Interest: The authors declare no conflict of interest.

References

1. FAO. Global Food Losses and food Waste. Extent, Causes and Prevention. 2011. Available online: https://www.fao.org/3/mb060e/mb060e00.pdf (accessed on 18 March 2023).
2. Onipe, O.O.; Ramashia, S.E.; Jideani, A.I. Wheat bran modifications for enhanced nutrition and functionality in selected food products. *Molecules* **2021**, *26*, 3918. [CrossRef] [PubMed]
3. Prueckler, M.; Siebenhandl-Ehn, S.; Apprich, S.; Hoeltinger, S.; Haas, C.; Schmid, E.; Kneifel, W. Wheat bran-based biorefinery 1: Composition of wheat bran and strategies of functionalization. *LWT-Food Sci. Technol.* **2014**, *56*, 211–221. [CrossRef]
4. Onipe, O.O.; Jideani, A.I.; Beswa, D. Composition and functionality of wheat bran and its application in some cereal food products. *Int. J. Food Sci. Technol.* **2015**, *50*, 2509–2518. [CrossRef]
5. Squeo, G.; Silletti, R.; Napoletano, G.; Greco Miani, M.; Difonzo, G.; Pasqualone, A.; Caponio, F. Characterization and Effect of Refining on the Oil Extracted from Durum Wheat By-Products. *Foods* **2022**, *11*, 683. [CrossRef]
6. Vurro, F.; Greco Miani, M.; Summo, C.; Caponio, F.; Pasqualone, A. Effect of durum wheat oil on the physico-chemical and sensory features of biscuits. *Foods* **2022**, *11*, 1282. [CrossRef]
7. Vurro, F.; Summo, C.; Squeo, G.; Caponio, F.; Pasqualone, A. The Use of Durum Wheat Oil in the Preparation of Focaccia: Effects on the Oxidative Stability and Physical and Sensorial Properties. *Foods* **2022**, *11*, 2679. [CrossRef]
8. Schutyser, M.A.I.; Van der Goot, A.J. The potential of dry fractionation processes for sustainable plant protein production. *Trends Food Sci Technol.* **2011**, *22*, 154–164. [CrossRef]
9. Bühler, J.M.; Schlangen, M.; Möller, A.C.; Bruins, M.E.; van der Goot, A.J. Starch in plant-based meat replacers: A new approach to using endogenous starch from cereals and legumes. *Starch-Stärke* **2022**, *74*, 2100157. [CrossRef]
10. Verardo, V.; Gómez-Caravaca, A.M.; Messia, M.C.; Marconi, E.; Caboni, M.F. Development of functional spaghetti enriched in bioactive compounds using barley coarse fraction obtained by air classification. *J. Agric. Food Chem.* **2011**, *59*, 9127–9134. [CrossRef]
11. Silventoinen, P.; Kortekangas, A.; Ercili-Cura, D.; Nordlund, E. Impact of ultra-fine milling and air classification on biochemical and techno-functional characteristics of wheat and rye bran. *Food Res. Int.* **2021**, *139*, 109971. [CrossRef]
12. Pojić, M.; Mišan, A.; Tiwari, B. Eco-innovative technologies for extraction of proteins for human consumption from renewable protein sources of plant origin. *Trends Food Sci. Technol.* **2018**, *75*, 93–104. [CrossRef]
13. Alcorta, A.; Porta, A.; Tárrega, A.; Alvarez, M.D.; Vaquero, M.P. Foods for plant-based diets: Challenges and innovations. *Foods* **2021**, *10*, 293. [CrossRef]
14. GFI, The Good Food Institute. U.S. Retail Market Insights–Plant-Based Foods. 2021. Available online: https://gfi.org/wp-content/uploads/2022/03/2021-U.S.-retail-market-insights_Plant-based-foods-GFI.pdf (accessed on 20 April 2023).
15. López-Martínez, M.I.; Miguel, M.; Garcés-Rimón, M. Protein and sport: Alternative sources and strategies for bioactive and sustainable sports nutrition. *Front. Nutr.* **2022**, *9*, 926043. [CrossRef]
16. Westberg, K.; Stavros, C.; Parker, L.; Powell, A.; Martin, D.M.; Worsley, A.; Fouvy, D. Promoting healthy eating in the community sport setting: A scoping review. *Health Promot. Int.* **2022**, *37*, 1–17. [CrossRef]
17. Arribas, C.; Cabellos, B.; Cuadrado, C.; Guillamón, E.; Pedrosa, M.M. Extrusion effect on proximate composition, starch and dietary fibre of ready-to-eat products based on rice fortified with carob fruit and bean. *LWT-Food Sci. Technol.* **2019**, *111*, 387–393. [CrossRef]
18. Gill, A.; Meena, G.S. Formulation of functional energy bars using dairy and non-dairy ingredients: A review. *Int. J. Chem. Stud.* **2020**, *8*, 1337–1342. [CrossRef]
19. Ciurzyńska, A.; Cieśluk, P.; Barwińska, M.; Marczak, W.; Ordyniak, A.; Lenart, A.; Janowicz, M. Eating habits and sustainable food production in the development of innovative "healthy" snacks. *Sustainability* **2019**, *11*, 2800. [CrossRef]
20. Anandito, R.B.K.; Nurhartadi, E.; Agustiani, R.B. Formulation of snack bars made from black rice bran (*Oryza sativa* L.) and sweet potato flour (*Ipomoea batatas* L.). *Conf. Ser. Earth Environ. Sci.* **2021**, *828*, 12028.
21. Asioli, D.; Aschemann-Witzel, J.; Caputo, V.; Vecchio, R.; Annunziata, A.; Næs, T.; Varela, P. Making sense of the "clean label" trends: A review of consumer food choice behavior and discussion of industry implications. *Food Res. Int.* **2017**, *99*, 58–71. [CrossRef]
22. Osborn, S. Labelling relating to natural ingredients and additives. In *Advances in Food and Beverage Labelling*; Woodhead Publishing: Sawston, UK, 2015; pp. 207–221.
23. Boukid, F.; Klerks, M.; Pellegrini, N.; Fogliano, V.; Sanchez-Siles, L.; Roman, S.; Vittadini, E. Current and emerging trends in cereal snack bars: Implications for new product development. *Int. J. Food Sci. Nutr.* **2022**, *73*, 610–629. [CrossRef]
24. Singh, A.; Kumari, A.; Chauhan, A.K. Formulation and evaluation of novel functional snack bar with amaranth, rolled oat, and unripened banana peel powder. *J. Food Sci. Technol.* **2022**, *59*, 3511–3521. [CrossRef] [PubMed]
25. Baurina, A.V.; Baurin, D.V.; Shakir, I.V.; Panfilov, V.I. Use of Sunflower Protein in Snack Bars. *Chem. Eng. Trans.* **2021**, *87*, 1–6.
26. De Sousa, M.F.; Guimarães, R.M.; de Oliveira Araújo, M.; Barcelos, K.R.; Carneiro, N.S.; Lima, D.S.; Egea, M.B. Characterization of corn (*Zea mays* L.) bran as a new food ingredient for snack bars. *LWT-Food Sci. Technol.* **2019**, *101*, 812–818. [CrossRef]
27. Meethal, S.M.; Kaur, N.; Singh, J.; Gat, Y. Effect of addition of jackfruit seed flour on nutrimental, phytochemical and sensory properties of snack bar. *Curr. Res. Nutr. Food Sci.* **2017**, *5*, 154–158. [CrossRef]
28. Squeo, G.; De Angelis, D.; Leardi, R.; Summo, C.; Caponio, F. Background, applications and issues of the experimental designs for mixture in the food sector. *Foods* **2021**, *10*, 1128. [CrossRef]

29. Leardi, R. D-Optimal Designs. In *Encyclopedia of Analytical Chemistry*; Meyers, R.A., Ed.; Wiley: Hoboken, NJ, USA, 2006; pp. 1–11.
30. AOAC. Official Methods of Analysis of AOAC International. In *AOAC International*, 18th ed.; AOAC: Gaithersburg, MD, USA, 2006.
31. Costantini, M.; Summo, C.; Faccia, M.; Caponio, F.; Pasqualone, A. Kabuli and Apulian black chickpea milling by-products as innovative ingredients to provide high levels of dietary fibre and bioactive compounds in gluten-free fresh pasta. *Molecules* **2021**, *26*, 4442. [CrossRef]
32. Pasqualone, A.; Makhlouf, F.Z.; Barkat, M.; Difonzo, G.; Summo, C.; Squeo, G.; Caponio, F. Effect of acorn flour on the physico-chemical and sensory properties of biscuits. *Heliyon* **2019**, *5*, e02242. [CrossRef]
33. Amigo, J.M.; Babamoradi, H.; Elcoroaristizabal, S. Hyperspectral image analysis. A tutorial. *Anal. Chim.* **2015**, *896*, 34–51. [CrossRef]
34. Leardi, R.; Melzi, C.; Polotti, G. CAT (Chemometric Agile Tool). 2013. Available online: http://gruppochemiometria.it/index.php/software (accessed on 20 April 2023).
35. Kim, E.J.; Corrigan, V.K.; Hedderley, D.I.; Motoi, L.; Wilson, A.J.; Morgenstern, M.P. Predicting the sensory texture of cereal snack bars using instrumental measurements. *J. Texture Stud.* **2009**, *40*, 457–481. [CrossRef]
36. Ciudad-Mulero, M.; Fernández-Ruiz, V.; Matallana-González, M.C.; Morales, P. Dietary fiber sources and human benefits: The case study of cereal and pseudocereals. *Adv. Food Nutr. Res.* **2019**, *90*, 83–134.
37. European Parliament. Regulation (EU) 1169/2011 of the European Parliament and of the Council of 25 October 2011 on the provision of food information to consumers. *Off. J. Eur. Comm.* **2011**, *L304*, 18–63.
38. EFSA. Sets European Dietary Reference Values for Nutrient Intakes. 2010. Available online: https://www.efsa.europa.eu/en/press/news/nda100326 (accessed on 25 June 2023).
39. Garcia, M.C.; Lobato, L.P.; Benassi, M.D.T.; Soares Júnior, M.S. Application of roasted rice bran in cereal bars. *Food Sci. Technol.* **2012**, *32*, 718–724. [CrossRef]
40. Zhubi-Bakija, F.; Bajraktari, G.; Bytyçi, I.; Mikhailidis, D.P.; Henein, M.Y.; Latkovskis, G.; Zirlik, A. The impact of type of dietary protein, animal versus vegetable, in modifying cardiometabolic risk factors: A position paper from the International Lipid Expert Panel (ILEP). *Clin. Nutr.* **2021**, *40*, 255–276. [CrossRef]
41. Camerotto, C.; Cupisti, A.; D'Alessandro, C.; Muzio, F.; Gallieni, M. Dietary fiber and gut microbiota in renal diets. *Nutrients* **2019**, *11*, 2149. [CrossRef]
42. European Parliament and the Council. Regulation (EC) No 1924/2006 of the European Parliament and of the Council of 20 December 2006 on nutrition and health claims made on foods. *Off. J. Eur. Union* **2006**, *L304*, 9–25.
43. Umme, H.; Ashadujjaman, R.M.; Mehedi, H.M.; Afroz, T.M.; Delara, A.; Rahman, M.M.A. Nutritional, textural, and sensory quality of bars enriched with banana flour and pumpkin seed flour. *Foods Raw Mater.* **2021**, *9*, 282–289.
44. Mathu, M.; Kumari, A. Processing & utilization of fruits and vegetable in cereal, legume and pseudo cereal based composite bars. *J. Pharm. Innov.* **2022**, *11*, 42–54.
45. Zulaikha, Y.; Yao, S.H.; Chang, Y.W. Physicochemical and Functional Properties of Snack Bars Enriched with Tilapia (*Oreochromis niloticus*) By-Product Powders. *Foods* **2021**, *10*, 1908. [CrossRef]
46. Miquel-Kergoat, S.; Azais-Braesco, V.; Burton-Freeman, B.; Hetherington, M.M. Effects of chewing on appetite, food intake and gut hormones: A systematic review and meta-analysis. *Physiol. Behav.* **2015**, *151*, 88–96. [CrossRef]
47. Silva, J.S.; Damiani, C.; Silva, E.P.; Ruffi, C.R.G.; Asquieri, E.R.; Vilas Boas, E.V.D.B. Effect of marolo (*Annona crassiflora* Mart.) pulp flour addition in food bars. *J. Food Qual.* **2018**, *2018*, 8639525. [CrossRef]
48. Arsa, S.; Theerakulkait, C. Preparation, aroma characteristics and volatile compounds of flavorings from enzymatic hydrolyzed rice bran protein concentrate. *J. Sci. Food Agric.* **2018**, *98*, 4479–4487. [CrossRef] [PubMed]

foods

MDPI

Article

Assessment of the Malting Process of *Purgatory* Bean and *Solco Dritto* Chickpea Seeds

Alessio Cimini, Alessandro Poliziani, Lorenzo Morgante and Mauro Moresi *

Dipartimento per l'Innovazione nei sistemi Biologici, Agroalimentari e Forestali, Università della Tuscia, Via S. C. de Lellis, 01100 Viterbo, Italy; a.cimini@unitus.it (A.C.); alessandro.poliziani@unitus.it (A.P.); lorenzo.morgante.98@gmail.com (L.M.)
* Correspondence: mmoresi@unitus.it; Tel.: +39-0761-357497

Abstract: This study was aimed at minimizing the anti-nutrient content of the Gradoli *Purgatory* bean (GPB: *Phaseolus vulgaris*) and *Solco Dritto* chickpea (SDC: *Cicer arietinum*) seeds grown in the Latium region of Italy by defining the three steps of their malting process. The water steeping and germination phases were carried out in a 1.0-kg bench-top plant at 18, 25, or 32 °C. By soaking both seeds at 25 °C for 3 h, 95 to 100% of seeds sprouted. There was no need for prolonging their germination process after 72 h, the degradation degree of raffinose in germinated GPBs or SDCs being about 63%, while that of phytic acid being ~32% or 23%, respectively. The steeping and germination kinetics of both seeds were mathematically described via the Peleg and first-order reaction models, respectively. The third step (kilning) was carried out under fluent dry air at 50 °C for 24 h and at 75 °C for 3 h, and yielded cream-colored malted seeds, the cotyledons of which were cyclonically separated from the cuticles and finally milled. Owing to their composition, the decorticated malted pulse flours might be used in the formulation of specific gluten-free food products high in raw proteins and low in phytate, α-oligosaccharides and in vitro glycemic index (GI). Even if their low GI trait was preserved after malting, only the GPB malt flour having a resistant starch-to-total starch ratio ≥ 14% has the potential to be labeled with the health claim for improving postprandial glucose metabolism according to EU Regulation 432/2012.

Keywords: decorticated malted pulses; germination kinetics; in vitro glycemic index; kilning conditions; pulse flour; pulse malting; pulse processing; raffinose and phytate removal; steeping kinetics

Citation: Cimini, A.; Poliziani, A.; Morgante, L.; Moresi, M. Assessment of the Malting Process of *Purgatory* Bean and *Solco Dritto* Chickpea Seeds. *Foods* **2023**, *12*, 3187. https://doi.org/10.3390/foods12173187

Academic Editors: Jennifer Ahn-Jarvis and Brittany A. Hazard

Received: 25 July 2023
Revised: 22 August 2023
Accepted: 22 August 2023
Published: 24 August 2023

1. Introduction

Legumes are rich in proteins, dietary fibers, and micronutrients and are currently used to prepare several pulse-based food products to replace foods of animal origin in the diet, as well as gluten-free products [1]. Despite their high nutritional profile [2] and sustainable farming [3], their global per capita consumption has stagnated during the last three decades and currently is no more than 21 g/day [4], probably because of their long cooking times, unpleasant flavor, low-digestible proteins, gastrointestinal problems [5], and high content of anti-nutrients (i.e., phytic acid, tannins, enzyme inhibitors, and flatulence-inducing oligosaccharides) [6].

In 2021, the worldwide production of pulses was nearly 89 million metric tons (Mg) [7]. Of this, the production of dry bean accounted for about 27.7 million Mg [8], chickpea production for about 14.3 million Mg [9], and lentil production for about 5.7 million Mg [10]. The largest world producer of dry beans and chickpeas is India with about 6.1 and 9.9 million Mg per year, respectively, while the second producer is Turkey for dry bean production (630×10^3 Mg) or Brazil for chickpea production (2.9×10^6 Mg). In Italy, dry pulse production decreased significantly from about 640,000 Mg in 1960 to 135,000 Mg in 2010 [11]. Even if the trend has started to revert since 2017, the import-to-consumption ratio is still as high as 98% for lentils, 95% for beans, and 59% for chickpeas.

Among the pulse varieties cultivated in Italy, three (i.e., *Gradoli Purgatory* beans, GPB; *Solco Dritto* chickpeas, SDC; and Onano lentils, OL) are typical of the Latium region [12]. In previous work [13], the lentils of Onano (Viterbo, Italy), which were awarded with the Protected Geographical Indication (PGI) mark by the European Union (PGI-IT-02651; https://www.tmdn.org/giview/gi/EUGI00000017728, accessed 6 August 2023), were submitted to malting and used to prepare fresh egg-pasta using only malted lentil flour. Such a fresh egg pasta was high in raw proteins (24 g/100 g), low in phytate (0.6 g/100 g) and in vitro glycemic index (41%), and basically free of oligosaccharides.

In this work, the other two pulse varieties were submitted to malting to attempt lowering their anti-nutrient content. In particular, the small, round, and whitish Purgatory bean seeds, similar to Cannellini beans but with a more delicate taste and a thinner skin, have been cultivated in the province of Viterbo (Italy), specifically in the towns of Gradoli, Acquapendente, and Onano, as testified by the traditional menu of the Purgatory Lunch, which has been served at Gradoli in the first day (Ash Wednesday) of the penitential Lenten season since the XVIIth century [14]. The other pulse variety accounted for the so-called *Solco Dritto* (straight furrow) chickpea (SDC) from the furrow tracing performed in the plain beneath the town of Valentano (Italy) on 14 August of every year, its straightness being regarded as a presage of an excellent harvest [15]. These smooth, yellow-beige skinned chickpea seeds have been locally cultivated since the time of Etruscans.

To minimize the anti-nutrient contents of pulses, various traditional (i.e., dehulling, soaking, boiling, pressure cooking, sprouting, and fermentation) and emerging (i.e., dielectric heating, extrusion, γ-irradiation, ultrasound, and high hydrostatic pressure) processing techniques have been tested with different reduction yields [16].

Malting, a conventional process used in the beer industry to obtain malted barley, was successfully applied to reduce the native content of phytic acid and raffinose in two lentil varieties, including the Onano ones [13].

The first aim of this work was to identify the most proper operating conditions of the three phases (i.e., seed steeping, germination, and kilning) of the malting process of Gradoli *Purgatory* beans and *Solco Dritto* chickpeas. The second aim was to describe mathematically the kinetic rate constants of the seed steeping and germination steps. Finally, the third one was to obtain dehulled malted pulse flours low in phytate and α-galactosides as novel ingredients in the formulation of several gluten-free pulse-based food products, such as fresh and dry pastas with low in vitro glycemic index.

2. Materials and Methods

2.1. Raw Materials

Two varieties of legumes were used in this work. The Gradoli *Purgatory* beans (*Phaseolus vulgaris*) and *Solco Dritto* chickpeas (*Cicer arietinum*) were produced and supplied by *Il Cerqueto* Srl (Acquapendente, Viterbo, Italy).

2.2. Physical Properties of Pulses

Both pulse seeds (as such or malted) were characterized by determining the mean values of the seed weight (m_S), volume (v_S), density (ρ_S), hydration capacity (HC), and swelling capacity (SC) by accounting for 50 seeds [17]. By assuming that each kernel had spherical conformation, it was possible to estimate the mean radius (R_S) of each seed [18].

2.3. Steeping Equipment

The steeping kinetics of each pulse seed was assessed in a bench-top plant, appropriately designed (Figure S1 in the electronic supplement). Each chamber was equipped with two stainless-steel perforated baskets, each one containing up to 1 kg of rehydrated seeds, and a low-temperature immersion circulator type IB-TastemakerCompact10 (Klarstein Chal-Tec GmbH, Berlin, Germany). As soon as the steeping process was completed, both stainless-steel baskets were moved to the germination chamber, where a sensor-type CJMCU-1080 HDC1080 (Texas Instruments, Dallas, TX, USA) was inserted to monitor the relative humidity (RH) and temperature (T) of the air with an accuracy of ±2% RH and ±0.2 °C, respectively. The germination chamber was thermostated at 18, 25, or 32 °C, while a temperature probe-type DS18B20 (Maxim Integrated, San Jose, CA, USA) was used to measure continuously the temperature of the seeds with an accuracy of ±0.5 °C. Water was sprayed inside each germination chamber for 1 min every hour, while the germinating seeds were aerated by manual mixing every 24 h to attempt homogenizing the distribution of water and disrupting the aggregates formed by root sprouting.

Soaking trials were carried out by charging each of the two baskets with about 0.3 kg of dry seeds, which were then submerged with deionized water at 18, 25, or 32 °C. Several soaked kernels were sequentially collected as time increased from 0 to 24 h, rapidly blotted on paper towels, and placed on the weighing plate of a Kern DAB 100-3 thermostatic scale (Kern&Sohn GmbH, Balingen, Germany) to be dried till constant weight at 110 °C for about 20 min.

2.4. Seed Germination

During each steeping test, forty seeds were collected at times ranging from 0 to 8 h and laid over an absorbent paper sheet, pre-soaked in 50 mL of deionized water, within a (20 cm × 14.5 cm × 2.5 cm) box. This was sealed and housed in a dark chamber thermostated at the same soaking temperature for 24 or 48 h. All the seeds with a manifest root, independently of its length, were counted and referred to the overall number of seeds accounted for. In this way, it was possible to assess the germination degree after 24 (G_{24}) or 48 (G_{48}) h, and thus identify the steeping time (t_{Sm}) and temperature (T_{Sm}), as well as the seed moisture content, associated with the minimum number of non-sprouted seeds. Finally, the seeds moistened at T_{Sm} for as long as t_{Sm} were drained and let germinate up to 24, 48, 72, or 96 h to detect the degradation of phytic acid and α-galactosides using the Phytic Acid and Raffinose/Sucrose/D-Glucose Assay Kits (Megazyme Ltd., Bray, Ireland), respectively.

2.5. Germinated Seed Kilning

Both germinated pulses were finally dehydrated at 50 °C for 24 h, and then at 75 °C for 3 h using the Nobel Pro 6 ventilated dryer (Vita 5, Gronsveld, The Netherlands), thus obtaining GPB and SDC malts. By using a portable color-measuring instrument mod. D25-PC2 (Hunterlab, Reston, VA, USA) with a diffuse (0/45°) illuminating viewing geometry, it was possible to assess the color of the split seeds in the CIELAB color space.

2.6. Dehulling and Grinding of Malted Pulses

Malted pulse seeds were manually submitted to slight abrasion before being aspirated via a cyclone separator using a suction system. Figure S2a shows the cyclone used, which was designed and produced using a 3D printer. In this way, it was possible to split malted GPB (Figure S2b) or SDC (Figure S2c) seeds into a cotyledon-rich fraction (Figure S2d or Figure S2e) and a cuticle-rich one. An electric stone mill (Mockmill 200, Wolfgang Mock, Otzberg, Germany) was used to convert each cotyledon-rich fraction into a decorticated GPB or SDC malt flour, its fineness being regulated at level 2 out of 10. Total starch (TS) and resistant starch (RS) contents in any malted pulse flour were assayed using the enzymatic kits by Megazyme Ltd. (Bray, Ireland). TS was assessed in raw samples, once dried and ground, while RS was tested in cooked ones.

2.7. Cooking of Pulse Seeds as Such or Malted

About 750 g of GPB or SDC seeds were weighted and transferred into a container. After adding tap water at 20 °C using a water-to-seed ratio of 4 g/g [19], the seeds were let to soak at 20 °C for 16 h. Once the soaking water had been drained, the moistened seeds were transferred into a stainless-steel pot containing 3 kg of tap water and let to cook with the lid closed at 98 °C for 60 or 90 min, respectively, using a 2-kW induction-plate hob (INDU, Melchioni Spa, Milan, Italy). About 750 g of decorticated malted GPBs or SDCs were cooked without presoaking at 98 °C for 45 or 60 min, respectively, with the pot initially filled with 3 kg of tap water at room temperature. The hob knob was set at the nominal power of 2 kW till the water started boiling, then it was shifted to 0.4 kW till the end of seed cooking. All the cooked pulses were recovered from the cooking water using a colander, cooled by running tap water for 90 s, and drained.

2.8. In Vitro Glycemic Index

In vitro digestion of the cooked pulses as such or malted was carried out using the procedure developed by Zou et al. [20]. All the tests were at least triplicated. By assaying the time course of the concentration of glucose released (C_G) during the simulated digestion using the *D*-Glucose Assay Procedure K-GLUC 07/11 (Megazyme Ltd., Bray, Ireland), it was possible to plot the so-called digestogram for each cooked sample. The area under the digestogram (AUC) was numerically calculated for a total digestion time of 180 min using the Trapezoidal Rule. The AUC values were referred to the corresponding area estimated for a reference product (i.e., white bread) [21] to calculate the percentage starch hydrolysis index (SHI), this being equal to 100% for the reference white bread. The empirical formula developed by Granfeldt et al. [22] allowed the in vitro glycemic index (GI) to be finally estimated:

$$GI = 8.198 + 0.862 \times SHI \quad (1)$$

2.9. Statistical Analysis of Data

All the tests were carried out at least 3 times to estimate the mean value (μ) and standard deviation (sd) of any parameter assayed. The Tukey Test at a probability level (*p*) of 0.05 was used to assess the statistical significance of any parameter difference. One-way analysis of variance (ANOVA) was also performed using SYSTAT v. 8.0 (SPSS Inc., Chicago, IL, USA, 1998).

3. Results and Discussion

3.1. Physical Properties

Table 1 shows the main chemico-physical properties of the pulse seeds under study.

Table 1. Main chemico-physical properties and CIELab coordinates (L*, a*, b*) of Gradoli *Purgatory* bean (GPB) and *Solco Dritto* chickpea (SDC) seeds as such or malted (M).

Parameter	GPB	MGPB	SDC	MSDC	Unit
Raw protein	22.7 ± 1.7^{a}	23.4 ± 2.1^{a}	22.3 ± 1.7^{a}	23.6 ± 1.9^{a}	g/100 g dm
Total Starch (TS)	33.81 ± 1.66^{b}	34.96 ± 0.19^{b}	46.8 ± 0.6^{a}	45.2 ± 2.0^{a}	g/100 g dm
Resistant Starch (RS)	23.59 ± 0.34^{a}	22.01 ± 1.82^{a}	1.77 ± 0.22^{b}	1.19 ± 0.43^{b}	g/100 g dm
Phytic Acid (PA)	1.15 ± 0.12^{a}	0.78 ± 0.13^{b}	1.15 ± 0.12^{a}	0.79 ± 0.09^{b}	g/100 g dm
Raffinose (R)	5.31 ± 0.28^{a}	1.95 ± 0.20^{c}	3.80 ± 0.15^{b}	1.65 ± 0.11^{c}	g/100 g dm
Seed weight (m_S)	0.167 ± 0.003^{c}	0.133 ± 0.002^{d}	0.302 ± 0.010^{a}	0.219 ± 0.002^{b}	g/seed
Seed volume (v_S)	0.123 ± 0.006^{c}	0.103 ± 0.000^{d}	0.233 ± 0.012^{a}	0.191 ± 0.000^{b}	cm^3/seed
Mean seed radius (R_S)	0.309 ± 0.005^{c}	0.290 ± 0.000^{d}	0.382 ± 0.006^{a}	0.358 ± 0.000^{b}	cm/seed
Seed density (ρ_S)	1.35 ± 0.03^{a}	1.30 ± 0.002^{b}	$1.30 \pm 0.09^{a,b}$	1.14 ± 0.01^{c}	g cm^3
Hydration capacity (HC)	0.149 ± 0.006^{c}	0.132 ± 0.003^{d}	0.284 ± 0.006^{a}	0.182 ± 0.010^{b}	g/seed
Swelling capacity (SC)	0.313 ± 0.012^{c}	0.368 ± 0.007^{b}	0.503 ± 0.015^{a}	0.369 ± 0.012^{b}	cm^3/seed
L*	$71.0 \pm 1.7^{b,c}$	$73.3 \pm 1.5^{a,b}$	69.5 ± 1.6^{c}	75.1 ± 1.8^{a}	-
a*	0.6 ± 0.5^{c}	0.01 ± 0.61^{c}	3.7 ± 0.5^{a}	2.3 ± 0.6^{b}	-
b*	15.6 ± 1.7^{b}	19.0 ± 1.9^{b}	27.0 ± 2.3^{a}	27.0 ± 1.3^{a}	-

In each row, values with the same letter have no significant difference at $p < 0.05$.

As concerning the Gradoli *Purgatory* beans, their crude protein, total starch, phytic acid, and raffinose contents on a dry matter basis were in line with those of the many bean varieties cultivated worldwide [4,23–25]. Raw beans are generally classified as very small, small, average, normal, or big size if their seed weight is smaller than 0.2 g, ranges from 0.2 to 0.3 g, from 0.3 to 0.4 g, from 0.4 to 0.5 g, or greater than 0.5 g [25]; GPBs, having an average weight of 0.167 g/seed, are of very small size with an equivalent spherical radius of 0.309 ± 0.005 cm. Their density (ρ_S) was in line with that of some white bean varieties grown in Tunisia [26]. The swelling capacity (SC) of any seed depends upon its hydration capacity (HC). High HC and SC values favor bean processing (e.g., soaking, germination, decortication, and fermentation) for either extracting active principles or removing anti-nutritional components [27]. Moreover, the higher the hydration capacity, the lower the bean cooking time and hardness will be. HC and SC were found to vary from 0.17 to 0.54 g/seed and from 0.16 to 0.50 cm^3/seed, respectively, in several genotypes of dry beans cultivated in Turkey [28] and Tunisia [26], probably because of the different seed size, coat thickness, and water absorption characteristics [29]. Common beans may contain from 0.4 to 16.1 g of α-galactosides [30] and from 0.3 to 2.9 g of phytic acid [31] per 100 g of dry mass. The raffinose equivalent content of GPBs was around 5.3 $\pm$ 0.3 g/100 g, while the phytic acid content amounted to 1.15 $\pm$ 0.12 g/100 g (Table 1).

As concerning the *Solco Dritto* chickpeas, their crude protein, phytic acid, and raffinose contents on a dry matter basis (Table 1) complied with those of several chickpea varieties [32,33]. The seed weight (m_S), volume (v_S), hydration (HC), and swelling (SC) capacities were slightly smaller than those of a few Sicilian strains [17], but greater than some Indian cultivars of the Desi and Kabuli types [34]. Its density (1.3 $\pm$ 0.1 g/cm^3) was like that of the Indian chickpeas, but greater than that (1.18 $\pm$ 0.15 g/cm^3) of the Sicilian seeds. Since also in the case of chickpeas, the swelling capacity and hydration capacities were related to the cooking time [35], it is highly likely that the cooking time of SDCs would be intermediate among those of the above Indian and Sicilian varieties. SDCs contained about 3.8 g of raffinose and from 1.15 g of phytic acid per 100 g of dry mass, values quite near to those (4.2 $\pm$ 0.7 and 1.21 $\pm$ 0.09 g/100 g dm, respectively) of other Kabuli chickpea seeds grown in Egypt [36]. Moreover, the phytic acid content of SDCs was in line with the range of levels (0.3–1.4 g/100 g dm) assayed in several chickpea varieties by Sparvoli et al. [31]. Finally, once split, the GPB and SDC seeds were characterized by the CIELab color coordinates shown in Table 1. The difference in the lightness (L*) between these raw seeds was not statistically significant at p = 0.05 (Table 1), but the raw SDCs exhibited greater red-green (a*) and yellow-blue (b*) components than those of raw GPBs. Nevertheless, both split seeds were of a light color quite near to the cream color in the Avery list [37].

3.2. Soaking Kinetics of Dry Legumes

Figure 1 shows the time course of the moisture ratio (M) at the soaking temperatures of 18, 25, or 32 °C for both the pulse varieties under study. Table S1 in the electronic supplement shows the mean values (μ) and standard deviations (sd) of the experimental moisture weight fraction (x_W) against soaking time. All isotherms were characterized by an initial quick increase for t varying from 0 to about 3 or 5 h in the case of GPB or SDC seeds, respectively. After that, M exhibited a slower growth up to the equilibrium moisture ratio.

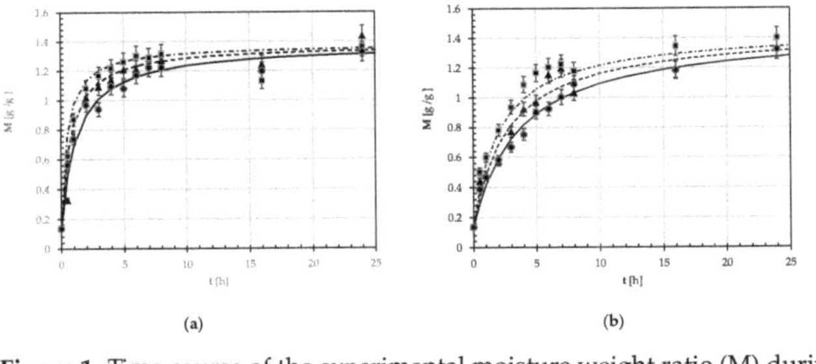

Figure 1. Time course of the experimental moisture weight ratio (M) during the steeping of Gradoli *Purgatory* beans (**a**) and *Solco Dritto* chickpeas (**b**) at different temperatures (•: —, 18 °C; ▲: - - -, 25 °C; ■, —. —, 32 °C). The continuous, broken, and dash-dotted line curves were plotted using the Peleg model (Equation (2)) and the constants k_1 and k_2 calculated as reported in the text.

The time course of the experimental moisture ratio (M) was described using the empirical model developed by Peleg [38]:

$$M(t) = M_0 + \frac{t}{k_1 + k_2 t} \tag{2}$$

where t is the soaking time, M_0 is the initial moisture ratio, while k_1 and k_2 are the Peleg rate and capacity constants.

The mean values and standard deviations of both Peleg constants were determined using the least squares method upon linearization of Equation (2):

$$\frac{t}{M(t) - M_0} = k_1 + k_2\, t \tag{3}$$

as shown in Table 2 together with the corresponding coefficient of determination (r^2).

Table 2. Mean values and standard deviations ($\mu \pm sd$) of Peleg constants k_1 and k_2 at different steeping temperatures (T) for Gradoli *Purgatory* beans and *Solco Dritto* chickpeas and empirical parameters [Ea/R; $ln(A)$] of the Arrhenius-type relationship (Equation (4)) and mean value $\left(\overline{k_2}\right)$.

Legume Variety Parameter	Gradoli *Purgatory* Beans			*Solco Dritto* Chickpeas			Unit
T	18	25	32	18	25	32	[°C]
k_1	0.92 ± 0.18	1.16 ± 0.29	0.36 ± 0.13	2.76 ± 0.26	2.23 ± 0.35	1.49 ± 0.14	[h g dm/g]
k_2	0.83 ± 0.02	0.75 ± 0.03	0.81 ± 0.01	0.75 ± 0.03	0.73 ± 0.04	0.71 ± 0.03	[g dm/g]
r^2	0.996	0.986	0.998	0.988	0.979	0.990	[-]
E_a/R	6099 ± 1278 [a]			3552 ± 441 [b]			[K]
$ln(A)$	−20.94 ± 4.29 [a]			−11.16 ± 1.48 [b]			[-]
r^2	0.85			0.93			[-]
$\overline{k_2}$	0.81 ± 0.02 [a]			0.76 ± 0.04 [a]			[g dm/g]

In each row, values with the same letter have no significant difference at $p < 0.05$.

Since the reciprocal of k_1 coincides with the initial water uptake rate (R_{W0}), its dependence on the absolute soaking temperature (T_K) was described using the following Arrhenius-type relationship:

$$\frac{1}{k_1} = A\, \exp\left(-\frac{E_a}{R\, T_K}\right) \tag{4}$$

where E_a is the activation energy, A is the pre-exponential nonthermal factor, and R (=8.31 J K^{-1} mol^{-1}) is the universal gas constant. The semilogarithmic plot shown in Figure 2 confirmed this assumption for both the pulses examined. On the contrary, the estimated k_2 values appeared to be about constant ($\overline{k_2}$) and temperature-independent, as also observed by other authors [39–41]. Table 2 also lists the least-squares estimated values of A and (E_a/R), as well as $\overline{k_2}$.

Finally, Figure 1 compares the experimental and calculated M values at the three temperatures examined for both pulses. For all the hydration isotherms examined, the average experimental error among the experimental and calculated M values was around 10 or 8% for the GPB or SDC seeds, respectively.

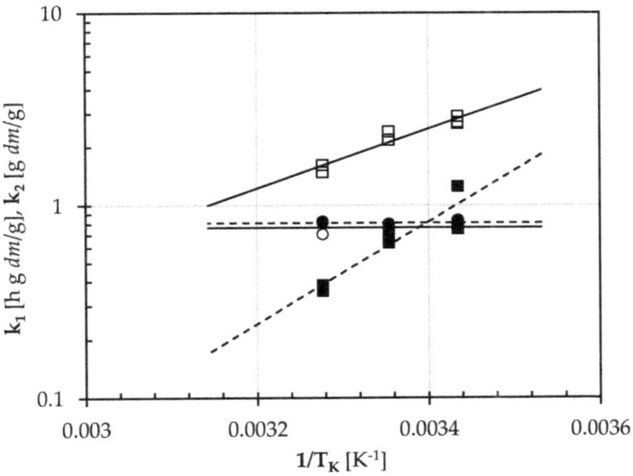

Figure 2. Peleg constants k_1 and k_2 of the soaking kinetics of Gradoli Purgatory beans (closed symbols and broken lines) and *Solco Dritto* chickpeas (open symbols and continuous lines): experimental and calculated k_1 (■, □) and k_2 (•, ○) values as a function of the reciprocal of the absolute soaking temperature (T_K). The broken and continuous lines fitting the k_1 values were plotted using Equation (4) with the parameters listed in Table 2, while the horizontal ones were plotted by averaging the k_2 values shown in Table 2.

For both the legumes under study, the independent parameters (A and E_a/R) were found to be statistically different at $p = 0.05$, while the Peleg capacity constant (k_2) was practically constant (0.78 ± 0.04 g dm/g) (Table 2).

As $t \to \infty$, the moisture ratio M approaches the equilibrium moisture content (M_e):

$$M_e = M_0 + \frac{1}{k_2} \tag{5}$$

The estimated value of the equilibrium moisture weight ratio (M_e) or fraction (x_{We}) for both legumes was approximately equal to 1.42 g/g dm or 58.6% *w/w*, respectively, in agreement with the values extrapolated from the data in Table S1.

3.3. Pulse Germinability

During the pulse seed soaking at 18, 25, or 32 °C, forty seeds were collected at times ranging from 0 to 8 h and let germinate at the same steeping temperature up to 24 or 48 h. Figure S3 shows the sealed boxes used to determine the soaking temperature and time associated with the maximum number of sprouted seeds. Table S2 reports the average number of GPBs and SDCs, exhibiting an evident root despite its longer or shorter length. By plotting the degree of germinability (G), that is the percentage of sprouted seeds out of 100 seeds, against the steeping time (t_S) (see Figure 3), it was noted that GPB seeds tended to sprout quicker than the SDC ones. However, after 48-h germination at 25 °C, 96 ± 4% of all the GPB seeds, but almost 100% of the SDC seeds sprouted after a presoaking time of 3 h. In such soaking conditions, which minimized the degree of inhomogeneity in germinating seeds, the average moisture content of the GPB or SDC seeds was equal to 52 ± 4 or 43 ± 3% (*w/w*), respectively (Table S1).

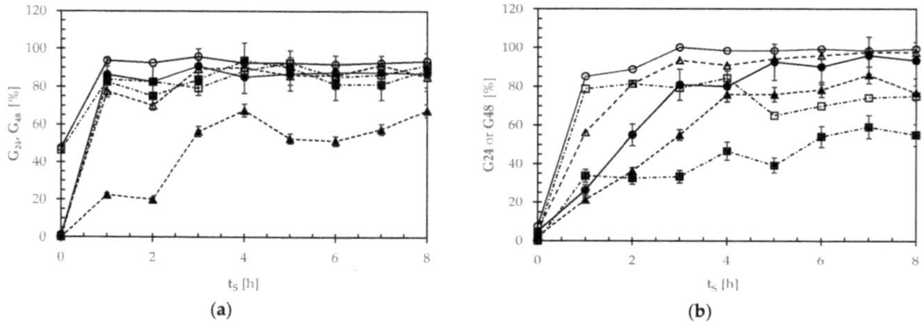

Figure 3. Average percentage of (**a**) Gradoli Purgatory beans and (**b**) *Solco Dritto* chickpeas germinated after 24 (G24: closed symbols) or 48 (G48: open symbols) h, once they had been presoaked at different temperatures (18 °C: ▲, △, - - -; 25 °C: •, ○, —; 32 °C: ■, □, —. —) for different steeping times (t_S).

3.4. Pulse Germination

Once the GPB and SDC seeds had been soaked at the above optimal conditions (T_{sm} = 25 °C, t_{sm} = 3 h), and the steeping water drained out, both rehydrated seeds were let to germinate at 25 °C up to 96 h.

Seed germination involves three steps, namely water imbibition, reactivation of metabolism, and radicle protrusion [42]. As seeds imbibe water, some physiological and biochemical processes (i.e., hydrolysis, macromolecules biosynthesis, respiration, subcellular structures, and cell elongation) are reactivated. These result in the hydrolysis of stored starch, polyphosphate, and other storage materials into simple forms. For instance, germinating seeds use sugars and other molecules as a substrate for respiration. The greatest storage form of total phosphorus (about 50–80%) in legumes is phytic acid ($C_6H_{18}O_{24}P_6$). This anti-nutrient may form complexes with proteins, and chelate some cations (i.e., Fe, Ca, K, Mn, Mg, Zn). The resulting mixed salts, such as phytin or phytate, in germinating seeds are hydrolyzed by an acid phosphatase enzyme (phytase), thus freeing phosphate, cations, and inositol easily utilizable by the seedlings [42].

Since the plot of the natural logarithm of the ratio between the current (C_i) and initial (C_{i0}) concentrations of the i-th anti-nutrient against the germination time (t_G) exhibited a linear negative trend for both the pulse seeds examined (Figure 4), the degradation kinetics of raffinose (R) or phytic acid (PA) were described as a first order reaction:

$$\frac{d\,C_i}{dt} = -\,k_i\,C_i \tag{6}$$

where k_i is the degradation kinetic rate constant of the i-th anti-nutrient. By separating the independent variables and integrating, the following was obtained:

$$C_i = C_{i0}\,e^{-k_i t} \qquad \text{for t} \geq 0 \tag{7}$$

where C_{i0} is the initial concentration of the i-th component.

Table 3. Mean values and standard deviations (μ ± sd) of the degradation kinetic rate constant k_i of each i-th component (R, PA) during the germination of Gradoli Purgatory beans and *Solco Dritto* chickpeas together with the corresponding coefficients of determination (r^2).

Parameter	k_i [h^{-1}]	r^2	k_i [h^{-1}]	r^2
Component i	Gradoli *Purgatory* beans		*Solco Dritto* chickpeas	
Raffinose	−0.018 ± 0.003 [a]	0.89	−0.012 ± 0.002 [b]	0.92
Phytic acid	−0.0044 ± 0.0005 [a]	0.96	−0.0035 ± 0.0003 [a]	0.97

In each row, values with the same letter have no significant difference at $p < 0.05$.

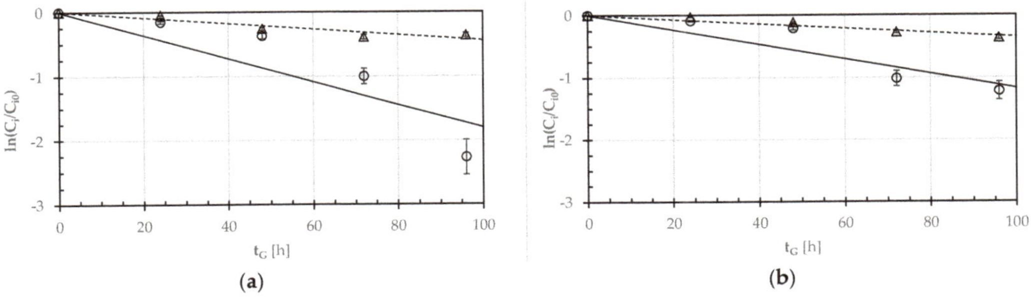

Figure 4. Semi-logarithmic diagram of the ratio (C_i/C_{i0}) for the raffinose (R: ○. —) and phytic acid (PA: △, - - - -) concentrations during the germination of GPBs (**a**) and SDCs (**b**) at 25 °C as a function of the germination time (t_G). The continuous and broken lines were plotted using the first-order kinetic model (Equation (6)) and the kinetic constant rates reported in Table 3.

Table 3 shows the least-squares estimates of k_i for both the pulses of concern.

It can be noted that the degradation rate constant of raffinose during the germination of GPBs was one and a half greater than that relative to the germinating SDCs, while the degradation rate constant of phytic acid for both seeds was not statistically different at $p < 0.05$.

Finally, Figure 5 shows the evolution of the concentration of raffinose and phytic acid during the germination of GPBs and SDCs at 25 °C. The lines plotted were calculated using Equation (7) with the kinetic constant rates shown in Table 3. It can be noted as quite a good reconstruction of the experimental profiles.

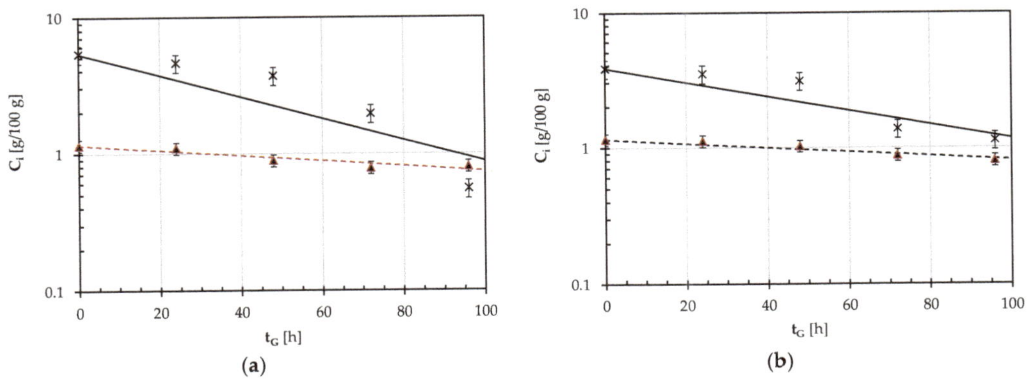

Figure 5. Germination of GPBs (**a**) and SDCs (**b**) at 25 °C: Concentration (C_i) of raffinose (R: ∗, —) or phytic acid (PA: ▲, - - -) against the germination time (t_G). The continuous and broken lines were plotted using the first-order kinetic model (Equation (7)) and the kinetic constant rates reported in Table 3.

In summary, there was no need for extending the seed germination process at 25 °C up to 96 h (Table S3). After 72-th, the degradation degree of raffinose or phytic acid in the germinating GPBs and SDCs was about 62% or 32%, and 63% or 23%, respectively. These phytate degradation degrees were smaller than those reported for a 96-h germination of Kabuli-type chickpeas (73%) and kidney bean from Ethiopia (79–96%) at 25 °C [43], as well as those measured in germinated black (53%) and white (45%) [44]. As concerning the only soaking in water of common beans and chickpeas, the reduction degree of raffinose or phytate was found to range from 22 to 97% or from 0.2 to 35%, respectively [16].

3.5. Malted Pulse Flour Production and Characterization

Once dehydrated firstly at 50 °C for 24 h and then at 75 °C for 3 h, the germinated pulse seeds gave rise to malted seeds having a moisture content of about 10% (*w*/*w*).

By using the cyclone shown in Figure S2a, almost all the cuticle fragments were separated from the cotyledons of both malted pulses. The cotyledon-rich fraction recovered approximately represented 85% or 86% of the input malted GPBs or SDCs, respectively.

As shown in Table 1, there was a general reduction in the physical properties of the decorticated malted pulses with respect to raw pulse ones. Moreover, the content of raffinose or phytate in malted GPBs or SDCs was reduced to about 37% or 68% of the original one, respectively.

Table 1 also shows the CIELab coordinates of the split malted seeds. The difference in their lightness L* was not statistically significant at $p = 0.05$, but the malted SDCs exhibited greater red-green (a*) and yellow-blue (b*) components than malted GPBs.

The difference in the CIELab coordinates (L*, a*, b*) between the raw and malted GPBs was not statistically significant, while the malted SDCs had greater lightness and a smaller red-green component (a*) than the raw counterpart, but the same yellow-blue component (b*). Since chickpea is a rich source of carotenoids, such as xanthophyll (9.0–19.7 mg/100 g), canthoxanthine (21.0–67.9 mg/100 g), and β-carotene (166–431 μg/100 g) [45], the significant increase in the lightness L* of malted SDCs may be highly likely attributed to the oxidative reactions, especially *cis-trans* isomerization of carotenoids, that occur during air kilning, these being intensified by higher temperature and lower relative humidity drying conditions [46]. Altogether, both malted seeds displayed a light color of the cream type in the Avery list [37].

Upon grinding for two or three cycles, it was possible to obtain a dehulled GPB or SDC malt flour, their raw protein, total starch, resistant starch, raffinose, and phytic acid contents coinciding with those of decorticated malted pulse seeds (Table 1).

Both these flours were characterized by almost the same raw protein (~23.5 g/100 g dm), raffinose (~1.8 g/100 g dm), and phytic acid (~0.78 g/100 g dm) contents, but quite different total and resistant starch concentrations. Of these, owing to its low total starch content (~35 g/100 g dm) and high resistant starch-to-total starch ratio (~0.63 g/g), the dehulled GPB malt flour might represent a valuable ingredient for the formulation of functional foods having a resistant starch level $\geq$ 14% of their TS content, this allowing their labelling with a specific health claim regarding the physiological effect of improved postprandial glucose metabolism according to EU Regulation 432/2012 [47].

3.6. In Vitro Glycemic Index of Pulse Seeds

To describe the simulated digestion kinetics of the cooked pulses as such or malted, the average concentration (C_G) of glucose freed by the enzymatic treatments was plotted against the incubation time (t), as shown in Figure 6.

Figure 6 shows that white bread exhibits a glycemic index significantly greater than the pulses, as also confirmed by the numerical calculation of the areas under each digestogram (AUC) up to an overall incubation time of 180 min (cfr. Table 4). It can be noted that AUC reduced from about 81 g min/L in the case of white bread to as low as 4.2 or 4.9 g min/L for the cooked SDC or GPBs seeds as such, the difference between these AUC values being not statistically significant at $p = 0.05$. For the malted GPB and SDC seeds, the AUC values were slightly higher (6.0 or 7.6 g min/L), but even their difference was statistically negligible at $p = 0.05$. Thus, by using Equation (1), the estimated in vitro glycemic index (GI) of both GPBs and SDCs as such or malted was equal to ~13 or 15%, respectively. Since such malted flours had a resistant starch content not statistically different from that of their native forms (Table 1), the in vitro tests shown in Figure 6 suggested that the malting process did not significantly change the GI response of the pulses.

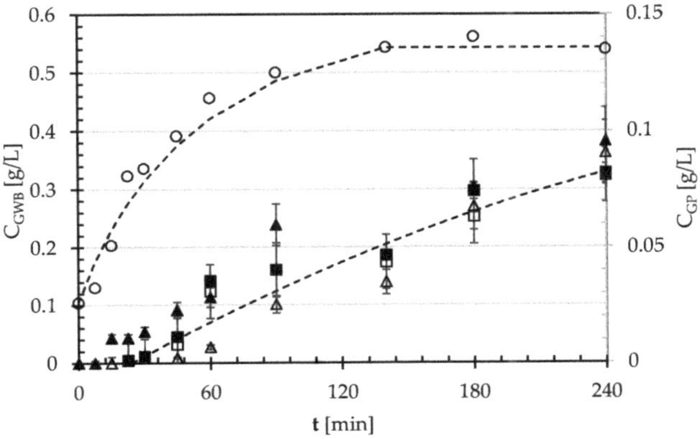

Figure 6. Glucose concentration freed by the simulated in vitro digestion of white bread (C_{GWB}: ○) or cooked pulses (C_{GP}) versus time (t): GPBs as such (□) or malted (■); SDCs as such (△) or malted (▲).

Table 4. Estimation of the areas (AUC) enclosed by the digestograms of white bread, Gradoli *Purgatory* beans as such (GPB) or malted (MGPB), and *Solco Dritto* chickpeas as such (SDC) or malted (MSDC) for a digestion time of 180 min using the Trapezoidal Rule, starch hydrolysis index (SHI), and in vitro glycemic index (GI) via Equation (1), and their classification according to the so-called GI chart.

Food Product	AUC [g min/L]	SHI [%]	GI [%]	GI Chart
White bread	81.2 ± 0.4 [a]	100.0 ± 0.5 [a]	94.4 ± 0.4 [a]	High
GPB as such	4.9 ± 0.2 [c]	6.0 ± 0.3 [c]	13.4 ± 0.2 [c]	Low
MGPB	6.0 ± 0.8 [b,c]	7.4 ± 1.0 [b,c]	14.6 ± 0.9 [b,c]	Low
SDC as such	4.2 ± 1.3 [c]	5.1 ± 1.5 [c]	12.6 ± 1.3 [c]	Low
MSDC	7.6 ± 1.1 [b]	9.3 ± 1.3 [b]	16.2 ± 1.1 [b]	Low

In each column, values with the same letter have no significant difference at $p < 0.05$.

Probably because of the different methods of preparation, processing, and heat application, the glycemic indexes reported in the literature [21] exhibit a wide variation from 18 ± 2 to $99 \pm 11\%$ for common beans, and from 14 ± 3 to $96 \pm 21\%$ for chickpeas when using white bread as a reference.

According to Foster-Powell et al. [48] and Atkinson et al. [49], foods can be classified into three categories: low (≤55), medium (55–69), and high (≥70) GI foods. Thus, the malting process did not affect the in vitro GI of the pulses examined here.

4. Conclusions

All the operating conditions of the three steps of the malting production process of two decorticated malted Gradoli *Purgatory* bean and *Solco Dritto* chickpea flours were defined together with the mathematical modelling of their seed steeping and germination processes. A three-h water steeping followed by a 72-h germination at 25 °C was sufficient to reduce the raffinose or phytic acid content by about 38% or 68–77% of the corresponding native content, respectively.

The subsequent kilning at 50 °C for 24 h and at 75 °C for 3 h gave rise to malted pulse seeds having almost the same cream color. The decorticated malted Gradoli *Purgatory* bean and *Solco Dritto* chickpea flours were characterized by quite different total and resistant starch concentrations, but almost the same raw protein (~23.5 g/100 g dm), raffinose (~1.8 g/100 g dm), and phytic acid (~0.78 g/100 g dm) contents, their low GI trait being preserved after malting. Thus, these flours might be regarded as valuable ingredients for designing several gluten-free food product formulations low in fats, α-oligosaccharide, and phytate specific for celiac, diabetic, and hyperlipidemic patients. However, just those prepared with the dehulled GPB malt flour (having a resistant starch level by far greater than 14% of its total starch content) might be labelled with the health claim of improved postprandial glucose metabolism according to EU Regulation 432/2012. Further work is needed to test the technical feasibility and sensory properties of such novel formulations.

Supplementary Materials: The following supporting information can be downloaded at: https://www.mdpi.com/article/10.3390/foods12173187/s1, Figure S1: Views of the experimental bench-top soaking chamber used; Figure S2: Picture of the laboratory-scale cyclone used to recover cotyledon-rich fractions; Figure S3: Pictures of the GPBs and SDCs during their germination in sealed boxes at different times; Table S1: Time course of the moisture weight fraction of GPBs and SDCs at different temperatures; Table S2: Average number of germinated GPBs and SDCs at different germination temperatures and times; Table S3: Effect of the germination time on the raffinose and phytic acid contents of GPBs and SDCs at 25 °C.

Author Contributions: Conceptualization, A.C. and M.M.; methodology, A.C. and M.M.; validation, A.C., A.P. and L.M.; formal analysis, A.C. and M.M.; investigation, A.C., A.P. and L.M.; resources, A.C. and M.M.; data curation, M.M.; writing—original draft preparation, M.M.; writing—review and editing, A.C., A.P., L.M. and M.M.; visualization and supervision: A.C. and M.M.; project administration, A.C.; funding acquisition, A.C. and M.M. All authors have read and agreed to the published version of the manuscript.

Funding: This research was supported by the GeCoWEB research project A0375-2020-3651 of the Lazio Region as part of the Public Notice "2020 Research Groups"—POR FESR Lazio 2014-2020—Action 1.2.1 approved with Resolution n. G08487 dated 19 July 2020—published in BURL No. 93 of 23 July 2020—amended with Resolution no. G10624/2020—published in BURL n. 116 of 22 September 2020.

Data Availability Statement: The data used to support the findings of this study can be made available by the corresponding author upon request.

Acknowledgments: The authors would like to thank the Giulia Longarini for her help during the experimental work.

Conflicts of Interest: The authors declare no conflict of interest.

References

1. Abu-Ghannam, N.; Gowen, A. Pulse-based food products. In *Pulse Foods: Processing, Quality and Nutraceutical Applications*, 2nd ed.; Tiwari, B.K., Gowen, A., McKenna, B., Eds.; Academic Press: London, UK, 2021; Chapter 15; pp. 369–391.
2. Maphosa, Y.; Jideani, V.A. The Role of Legumes in Human Nutrition. In *Functional Food—Improve Health through Adequate Food*; Chávarri Hueda, M., Ed.; IntechOpen Ltd.: London, UK, 2017; Chapter 6; pp. 103–121. Available online: https://www.intechopen.com/chapters/55808 (accessed on 23 August 2023).
3. Nemecek, T.; von Richthofen, J.-S.; Dubois, G.; Casta, P.; Charles, R.; Pahl, H. Environmental impacts of introducing grain legumes into European crop rotations. *Eur. J. Agron.* **2008**, *28*, 380–393. [CrossRef]
4. Rawal, V.; Navarro, D.K. *The Global Economy of Pulses*; FAO: Rome, Italy, 2019.
5. de Almeida Costa, G.E.; Da Silva Queiroz-Monici, K.; Pissini Machado Reis, S.M.; De Oliveira, A.C. Chemical composition, dietary fibre and resistant starch contents of raw and cooked pea, common bean, chickpea and lentil legumes. *Food Chem.* **2006**, *94*, 327–330. [CrossRef]
6. Gebrelibanos, M.; Tesfaye, D.; Raghavendra, Y.; Sintayeyu, B. Nutritional and health implications of legumes. *Int. J. Pharm. Sci. Res.* **2013**, *4*, 1269–1279.
7. Shahbandeh, M. Production Volume of Pulses Worldwide from 2010 to 2021. Statista, 2023. Available online: https://www.statista.com/statistics/721945/pulses-production-volume-worldwide/#statisticContainer (accessed on 25 June 2023).
8. OurWorldinData. Beans, Dry—Production (Tonnes). Available online: https://ourworldindata.org/grapher/bean-production?tab=table (accessed on 26 June 2023).
9. Atlasbig. World Chickpea Production by Country. Available online: https://www.atlasbig.com/en-us/countries-chickpea-production (accessed on 26 June 2023).
10. Atlasbig. World Lentil Production by Country. Available online: https://www.atlasbig.com/en-us/countries-lentil-production (accessed on 26 June 2023).
11. Confimi Industria. La Produzione di Legumi in Italia Riprende a Crescere. 2018. Available online: https://www.confimi.it/pagina-iniziale/868-la-produzione-di-legumi-in-italia-riprende-a-crescere (accessed on 5 July 2023).

12. Di Giovannantonio, C.; Catta, M.; Pica, G.; Casadei, G. *Lazio Patrimonio Agroalimentare tra Biodiversità e Tradizione*; ARSIAL: Rome, Italy, 2019.
13. Cimini, A.; Poliziani, A.; Morgante, L.; Moresi, M. Antinutrient removal in yellow lentils by malting. *J. Sci. Food Agric.* 2023; ahead of print. [CrossRef]
14. Slow Food Foundation. Purgatory Beans. Available online: https://www.fondazioneslowfood.com/en/ark-of-taste-slow-food/purgatory-beans/ (accessed on 27 June 2023).
15. Slow Food Foundation. Solco Dritto Chickpea. Available online: https://www.fondazioneslowfood.com/en/ark-of-taste-slow-food/solco-dritto-chickpea/ (accessed on 27 June 2023).
16. Das, G.; Sharma, A.; Sarkar, P.K. Conventional and emerging processing techniques for the post-harvest reduction of antinutrients in edible legumes. *Appl. Food Res.* **2022**, *2*, 100112. [CrossRef]
17. Patane, C.; Iacoponi, E.; Raccuia, S.A. Physico-chemical characteristics, water absorption, soaking and cooking properties of some Sicilian populations of chickpea (*Cicer arietinum* L.). *Int. J. Food Sci. Nutr.* **2004**, *55*, 547–554. [CrossRef]
18. Sayar, S.; Turhan, M.; Gunasekaran, S. Analysis of chickpea soaking by simultaneous water transfer and water-starch reaction. *J. Food Eng.* **2001**, *50*, 91–98. [CrossRef]
19. Wang, N.; Panozzo, J.F.; Wood, J.; Malcolmson, L.J.; Arganosa, G.C.; Baik, B.-K.; Driedger, D.; Han, J. AACCI approved methods Technical Committee report: Collaborative study on a method for determining firmness of cooked pulses (AACC International Method 56-36.01). *Cereal Foods World* **2012**, *57*, 230–234. [CrossRef]
20. Zou, W.; Sissons, M.; Gidley, M.J.; Gilbert, R.G.; Warren, F.J. Combined techniques for characterising pasta structure reveals how the gluten network slows enzymc digestion rate. *Food Chem.* **2015**, *188*, 559–568. [CrossRef]
21. Singh, M.; Manickavasagan, A.; Shobana, S.; Mohan, V. Glycemic index of pulses and pulse-based products: A review. *Crit Rev Food Sci Nutr.* **2021**, *61*, 1567–1588. [CrossRef]
22. Granfeldt, Y.; Björck, I.; Drews, A.; Towar, J. An in vitro procedure based on chewing to predict metabolic responses to starch in cereal and legume products. *Eur. J. Clin. Nutr.* **1992**, *46*, 649–660. [CrossRef]
23. Basso Los, F.G.; Ferreira Zielinski, A.A.; Wojeicchowski, J.P.; Nogueira, A.; Mottin Demiate, I. Beans (*Phaseolus vulgaris* L.): Whole seeds with complex chemical composition. *Curr. Opin. Food Sci.* **2018**, *19*, 63–71.
24. Cappa, C.; James, D.; Kelly, J.D.; Nga, P.K.W. Seed characteristics and physicochemical properties of powders of 25 edible dry bean varieties. *Food Chem.* **2018**, *253*, 305–313. [CrossRef] [PubMed]
25. de Barros, M.; Prudencio, S.H. Physical and chemical characteristics of common bean varieties. *Ciências Agrárias Londrina* **2016**, *37*, 751–762. [CrossRef]
26. Nciri, N.; El Mhamdi, F.; Ben Ismail, H.B.; Ben Mansour, A.; Fennira, F. Physical properties of three white bean varieties (*Phaseolus vulgaris* L.) grown in Tunisia. *J. Appl. Sci. Agric.* **2014**, *9*, 195–200.
27. Shimelis, E.A.; Rakshit, S.K. Proximate composition and physico-chemical properties of improved dry bean (*Phaseolus vulgaris* L.) varieties grown in Ethiopia. *LWT-Food Sci. Technol.* **2005**, *38*, 331–338. [CrossRef]
28. Ercan, R.; Atli, A.; Köksel, H.; Dağ, A. Cooking quality and composition of dry beans grown in Turkey. *GIDA* **1994**, *19*, 313–316.
29. Sefa-Dedeh, S.; Stanley, D.W. Textural implications of the microstructure of legumes. *Food Technol.* **1979**, *33*, 77–83.
30. Pedrosa, M.M.; Guillamón, E.; Arribas, C. Autoclaved and extruded legumes as a source of bioactive phytochemicals: A review. *Foods* **2021**, *10*, 379. [CrossRef]
31. Sparvoli, F.; Bollini, R.; Cominelli, E. Nutritional Value. In *Grain Legumes*; Ron, A.M.D., Ed.; Springer: New Work, NY, USA, 2015; pp. 291–326.
32. Frias, J.; Vidal-Valverde, C.; Sotomayor, C.; Diaz-Pollan, C.; Urbano, G. Influence of processing on available carbohydrate content and antinutritional factors of chickpeas. *Eur. Food Res. Technol.* **2000**, *210*, 340–345. [CrossRef]
33. Rawal, V.; Bansal, P.; Tyagi, K. Chickpea: Transformation in production conditions. In *The Global Economy of Pulses*; Rawal, V., Navarro, D.K., Eds.; FAO: Rome, Italy, 2019; Chapter 3; pp. 21–36.
34. Kaur, M.; Singh, N.; Singh Sodhi, N. Physicochemical, cooking, textural and roasting characteristics of chickpea (*Cicer arietinum* L.) cultivars. *J. Food Eng.* **2005**, *69*, 511–517. [CrossRef]
35. Williams, P.C.; Nakoul, H.; Singh, K.B. Relationship between cooking time and some physical characteristics in chickpeas (*Cicer arietinum* L.). *J. Sci. Food Agric.* **1983**, *34*, 492–496. [CrossRef]
36. El-Adawy, T.A. Nutritional composition and antinutritional factors of chickpeas (*Cicer arietinum* L.) undergoing different cooking methods and germination. *Plant Foods Hum. Nutr.* **2002**, *57*, 83–97. [CrossRef] [PubMed]
37. Converting Colors. Available online: https://convertingcolors.com/ (accessed on 12 August 2023).
38. Peleg, M. An empirical model for desorption of moisture sorption curve. *J. Food Sci.* **1988**, *53*, 1216–1219. [CrossRef]
39. Abu-Ghannam, N.; McKenna, B. Hydration kinetics of red kidney beans (*Phaseolus vulgaris* L.). *J. Food Sci.* **1997**, *62*, 520–523. [CrossRef]
40. Hung, T.V.; Liu, L.H.; Black, R.G.; Trewhella, M.A. Water absorption in chickpea (*C. arietinum*) and field pea (*P. sativum*) cultivars using the Peleg model. *J. Food Sci.* **2006**, *58*, 848–852. [CrossRef]
41. Pan, Z.; Tangratanavalee, W. Characteristics of soybeans as affected by soaking conditions. *Lebensm. Wiss. U.-Technol.* **2003**, *36*, 143–151. [CrossRef]
42. Ali, A.S.; Elozeiri, A.A. Metabolic Processes During Seed Germination. In *Seed Biology*; Jimenez-Lopez, J.C., Ed.; IntechOpen Ltd.: London, UK, 2017; Chapter 8; pp. 141–166. [CrossRef]

43. Haileslassie, H.A.; Henry, C.J.; Tyler, R.T. Impact of household food processing strategies on antinutrient (phytate, tannin and polyphenol) contents of chickpeas (*Cicer arietinum* L.) and beans (*Phaseolus vulgaris* L.): A review. *Int. J. Food Sci. Technol.* **2016**, *51*, 947–1957. [CrossRef]
44. Sangronis, E.; Machado, C.J. Influence of germination on the nutritional quality of *Phaseolus vulgaris* and *Cajanus cajan*. *LWT* **2007**, *40*, 116–120. [CrossRef]
45. Thavarajah, D.; Thavarajah, P. Evaluation of chickpea (*Cicer arietinum* L.) micronutrient composition: Biofortification opportunities to combat global micronutrient malnutrition. *Food Res. Int.* **2012**, *49*, 99–104. [CrossRef]
46. Rodriguez-Amaya, D.B. Isomerization and oxidation. In *Food Carotenoids: Chemistry, Biology, and Technology*; Rodriguez-Amaya, D.B., Ed.; John Wiley & Sons Ltd.: Chichester, UK, 2016; Chapter 7; pp. 174–198.
47. Commission Regulation (EU) No. 432/2012 of 16 May 2012 "Establishing a List of Permitted Health Claims Made on Foods, Other Than Those Referring to the Reduction of Disease Risk and to Children's Development and Health". *Off. J. Eur. Union* **2012**, *L136*, 1–40. Available online: https://eur-lex.europa.eu/legal-content/EN/TXT/PDF/?uri=CELEX:02012R0432-20170822 (accessed on 6 August 2023).
48. Foster-Powell, K.; Holt, S.H.A.; Brand-Miller, J.C. International table of glycemic index and glycemic load values. *Am. J. Clin. Nutr.* **2002**, *76*, 5–56. [CrossRef] [PubMed]
49. Atkinson, F.S.; Brand-Miller, J.C.; Foster-Powell, K.; Buyken, A.E.; Goletzke, J. International tables of glycemic index and glycemic load values 2021: A systematic review. *Am. J. Clin. Nutr.* **2021**, *114*, 1625–1632. [CrossRef] [PubMed]

foods

MDPI

Article

Fermentation of Peanut Slurry with *Lactococcus lactis* Species, *Leuconostoc* and *Propionibacterium freudenreichii* subsp. *globosum* Enhanced Protein Digestibility

Ayana Saizen, Letitia Stipkovits, Yukiyo Muto and Luca Serventi *

Faculty of Agriculture and Life Sciences, Lincoln University, Lincoln 7647, New Zealand
* Correspondence: luca.serventi@lincoln.ac.nz

Abstract: Peanuts contain nutritionally relevant levels of protein, yet are poorly digestible. Fermentation is a promising technique to boost legume protein quality, but its effect on the protein quality of raw peanuts has not been investigated. This study aimed to assess the impact of fermentation on the in vitro protein digestibility and free amino acid profile of cooked peanut slurry (peanut to water ratio 1:1). Cultures used were *Propionibacterium freudenreichii* subsp. *globosum* and a commercial fresh cheese culture that contained *Lactococcus lactis* subsp. *cremoris*, *lactis*, *lactis biovar diacetylactis*, and *Leuconostoc*, fermenting at 38 °C for 48 h. Samples fermented with the combination of cultures showed higher protein digestibility, as well as softer texture. Significant increases were observed only in the sample fermented with the fresh cheese culture. While the fresh cheese culture improved the free amino acid profile after fermentation, the combination of the cultures decreased all free amino acid concentrations except for glutamine, alanine, and proline. The observed increases in in vitro protein digestibility and the free amino acid profile may be attributed to the proteolytic activities of the cultures.

Keywords: protein digestibility; peanuts; legumes; probiotics; fermented food; *Lactococcus*; *Propionibacterium*; cheese; amino acids; texture

Citation: Saizen, A.; Stipkovits, L.; Muto, Y.; Serventi, L. Fermentation of Peanut Slurry with *Lactococcus lactis* Species, *Leuconostoc* and *Propionibacterium freudenreichii* subsp. *globosum* Enhanced Protein Digestibility. *Foods* **2023**, *12*, 3447. https://doi.org/10.3390/foods12183447

Academic Editors: Jennifer Ahn-Jarvis and Brittany A. Hazard

Received: 13 July 2023
Revised: 14 September 2023
Accepted: 14 September 2023
Published: 15 September 2023

1. Introduction

Legumes are one of the main sources of plant-based protein. They are rich in protein, fiber, carbohydrates, and micronutrients [1]. For instance, commercial raw peanuts contain 26.3 g of protein, 11.4 g of carbohydrates, and 5.7 g of dietary fiber per 100 g [2]. While legumes are nutritious foods, plant-based proteins generally have a lower digestibility rate (75 to 80%) compared to animal proteins (90 to 95%) [3]. Additionally, most plant-based proteins are incomplete sources of amino acids, while animal-based proteins tend to be complete sources [4]. In a previous study, it is reported that raw peanuts contained 270 ± 4 g/kg dry weight of protein which mainly consisted of glutamic acid (859 ± 23 mg/kg dry weight), phenylalanine (555 ± 84 mg/kg dry weight), asparagine (238 ± 79 mg/kg dry weight), valine (208 ± 9 mg/kg dry weight), aspartic acid (190 ± 8 mg/kg dry weight), arginine (188 ± 18 mg/kg dry weight), and alanine (170 ± 20 mg/kg dry weight) [5].

The protein quality of legumes can be enhanced through the fermentation technique. The quality of dietary protein is assessed based on its digestibility and amino acid profile [6]. Protein digestibility refers to the amount of hydrolyzed proteins that are broken down by digestive enzymes in relation to the total protein content [3]. Small peptide fractions can contribute to an increase in protein digestibility and accessibility in the body [7]. On the other hand, amino acid profiles can be expressed as free amino acid or essential amino acid profiles [3,8]. Ketnawa and Ogawa [8] discovered that soybeans fermented with *Bacillus natto* increased the protein digestibility with low molecular weight soluble-protein fractions. They also found that the fermentation technique increased the concentration of all free amino acids.

Although fermentation is a promising technique to boost legume protein quality, there is a lack of understanding of how it impacts the protein quality of raw peanuts. It was reported that the incorporation of naturally fermented peanuts into pearl millet-based infant foods improved the in vitro protein digestibility compared to the incorporation of naturally fermented cowpea [9]. It is also reported that sorghum base tempeh fermented with *Rhizopus oligosporus* increased the in vitro protein digestibility when mungbean and peanuts were incorporated [10]. However, there is no published research on the effect of fermentation with cheese cultures on the protein digestibility and free amino acid profile of raw peanuts. The aim of this research was to evaluate the effect of fermentation using a commercial fresh cheese culture and *Propionibacterium freudenreichii* subsp. *globosum* on the protein digestibility and free amino acid profile of raw peanut slurry. The fresh cheese culture contained *Lactococcus lactis* subspecies. The hypothesis posited that proteolysis activities of microbes would increase the protein digestibility and free amino acid profile of raw peanut slurry through fermentation.

2. Materials and Methods

2.1. Materials and Slurry Fermentation

Raw peanuts (Pams, Auckland, New Zealand) were deskinned and blended with water at a 1:1 (*w*/*w*) ratio until a smooth slurry formed. Fresh cheese culture was obtained from Mad Millie (Auckland, New Zealand); it contained *Lactococcus lactis* subsp. *cremoris, lactis, lactis biovar diacetylactis*, and *Leuconostoc*. *Propionibacterium freudenreichii* subsp. *globosum* culture was obtained from the Urban Cheese Company (West Melton, Christchurch, New Zealand). The fresh cheese culture was dissolved in water at a concentration of 0.10%. *Propionibacterium freudenreichii* subsp. *globosum* was also dissolved into the fresh cheese culture solution at a concentration of 0.03%. A baking pan was sanitized using 70% ethanol, and then 90 g of the raw peanut slurry was transferred to a cup on the pan. Subsequently, 32 g of each culture solution was added to the cup and mixed in. At this point, pre-fermentation samples were stored in a freezer for subsequent analysis. The samples were incubated at 38 °C for 48 h, then stored in a freezer for further analysis and thawed at room temperature prior to analysis. Each sample was prepared in triplicate. A representative picture of the slurry produced is available in Figure 1.

Figure 1. Representative picture of the fermented slurry produced from peanuts.

2.2. pH

One of the parameters used to test fermentability is pH. The pH of samples was measured with a Mettler Toledo pH meter (SevenEasy pH, Schwerzenbach, Switzerland). Slurries were analyzed in their original state by immersing the pH probe in them.

2.3. Microbial Enumeration

Fermentability was also tested by quantifying *Lactococcus* microorganisms. Duplicate sets of De Man–Rogosa–Sharpe (MRS) agar plates, pH 5.7, were labeled and inoculated with 1 mL of diluted suspensions, distributed with a spreader. The MRS media was chosen as ideal substrate for probiotic growth, specifically for lactobacilli such as *Lactococcus* and *Leuconostoc* [11]. Incubation took place at 35 °C for 48 h. Anaerobic sachets (Oxoid AnaeroGen 2.5 L, Thermo Scientific^{TM}; Christchurch, New Zealand) were used. Microbial count was expressed as colony forming units per gram of sample (CFU/g).

2.4. Protein Digestibility

2.4.1. Preparation of Sample Solutions

Aliquots of 1 g of raw peanut slurry were weighed and left overnight at room temperature in a 200 mL beaker. Then, 90 mL of distilled water was added to the 200 mL beaker to dissolve the slurry, and the mixture was stirred using a magnetic stirring bar until the slurry broke down into small particles. The pH of the solution was adjusted to 2.0 with 1 mol/L HCl, and then the volume was adjusted to 100 mL with distilled water, for a final pH of 2.0.

2.4.2. Preparation of Enzymes

Pepsin solution was prepared with 25 mg of pepsin (600–800 U/mg protein; EC 232-623-3) and 10 mL of 0.04 mol/L HCl for gastric digestion. Pancreatin solution was prepared with 25 mg of pancreatin (activity > 25 USP units/mg; CAS 8049-47-6) and 10 mL of 0.1 mol/L phosphate buffer for intestinal digestion combined with lipase (activity > 2.0 USP units/mg) and protease (activity > 25 USP units/mg).

2.4.3. In Vitro Protein Digestion

In vitro digestion was determined according to the methods of Uraipong and Zhao [12] with slight adjustments. Sample solutions and blank water were heated to 37 °C using magnetic stirrers (IKA, Guangzhou, China). Then, 1 mL aliquots were taken, followed by the addition of 0.5 mL of the pepsin solution to initiate the gastric phase. Samples were kept at 37 °C and stirred, and 1 mL of aliquots were taken every 20 min up to 120 min. The aliquots were heated in a 95 °C water bath for 10 min to quench the enzymatic reaction at each sampling point. After 120 min, the pH was adjusted to 8.0 with 1.7 mol/L NaOH and 1 mol/L HCl. The addition of 0.5 mL pancreatin solution initiated the intestinal phase, while the samples continued to be stirred and maintained at 37 °C. The 1 mL aliquots were again taken every 20 min for the subsequent 120 min, and the aliquots were treated as in the gastric phase to quench the enzymatic reaction at each sampling point. All in vitro digestion samples were stored at −20 °C for further analysis.

2.4.4. In-Vitro Protein Digestibility Evaluation

Protein concentration in the in vitro digestion samples was determined by the Bradford method. BSA solution (0.5 mg/mL) was diluted to form a standard curve (0, 0.125, 0.25, 0.4, and 0.5 mg/mL) using distilled water. Bio-Rad protein assay dye reagent was diluted at a reagent and water ratio of 1:4 (*v*/*v*), and then the diluted reagent was filtered. Then, 10 µL of each in vitro digestion sample and 200 µL of the diluted dye reagent were added to a cell on 96-well plates and then mixed three times using pipetting. The absorbance of the plates was measured at 595 nm within 1 h after mixing the first sample. As shown in Equation (1), the in vitro protein digestibility was calculated using the equation established by Almeida and collaborators [13]. In this equation, *Ph* represents the protein concentration in the in vitro digestion sample, *Pb* is the protein concentration in the blank, and *P*0 represents the protein concentration in the in vitro digestion sample at 0 min.

$$Protein\ Digestibility(\%) = \left(1 - \frac{Ph - Pb}{P0}\right) \times 100 \quad (1)$$

2.5. Total Amino Acids

Wet samples were defatted with the Soxhlet method. Defatted samples were then treated prior to chromatographic analysis. The amino acid profile of the freeze-dried sample was determined upon acid hydrolysis (5.0 mL 6 N HCl solutions heated at 110 °C for 20 h) and subsequent chromatographic analysis (Agilent 1100 Series HPLC system; Santa Clara, CA, USA). The HPLC settings were as follows: 150 × 4.6 mm, C18, 3u ACE-111-1546 column (Winlab, Harborough, UK) at a temperature of 40 °C. Two solvents were used: A (0.01 M disodium phosphate in 0.8% tetrahydrofuran, pH 7.5) and B (50% methanol, 50% acetonitrile) (LiChrosolv Reag, VWR, Radnor, PA, USA). The flow rate was 0.7 mL/min with solvent B increasing from 0 to 100% in 24 min, then decreasing to 0% in 12 min. Derivatization was performed with o-phthalaldehyde and 3-Mercaptopropionic acid for primary amino acids and 9-fluorenylmethyl chloroformate for secondary amino acids. Injection volume was 11.0 µL. A fluorescence detector (excitation 335 nm, emission 440 nm) was used. At 22 min, the detector was switched for secondary amino acid (excitation 260 nm, emission 315 nm).

2.6. Free Amino Acids

The quantity of free amino acids in wet samples was determined with the same chromatographic method described in Section 2.5, but without acid hydrolysis, as detailed elsewhere [14].

2.7. Texture Profile Analysis

Digestibility of protein is also affected by the texture and structure. Therefore, Texture Profile Analysis (TPA) was performed on peanut slurry. A TA.XT Texture Analyser (Stable Micro Systems, Godalming, UK) was used to perform double compression on slurry samples to a compression rate of 40%. Slurry was analyzed in the baking pan and settings were the following: load cell 50 kg, aluminum probe P/25, and test speed 1.7 mm/s. Parameters measured were hardness and adhesiveness.

2.8. Statistical Analysis

Values were repeated in duplicate (microbial enumeration, total and free ammino acids) and triplicate (pH, protein digestibility, hardness and adhesiveness). All calculations were performed using Excel from Microsoft Office Home and Business 2019, and Minitab (version 20) was utilized for statistical analysis. Results are reported as mean plus/minus standard deviation. Statistically significant differences were evaluated with one-way analysis of variance (ANOVA) and Tukey's honest significant difference (HSD) test ($p < 0.05$).

3. Results

3.1. Fermentability

As shown in Table 1, the pH of the raw peanut slurry decreased significantly after fermentation. The initial pH was 6.78 in the sample with the fresh cheese culture, and 6.67 in the sample with the combination of the fresh cheese culture and *Propionibacterium freudenreichii* subsp. *globosum* (Figure 2). Although the pH significantly decreased to 4.74 and 4.70 after fermentation, no significant difference was observed between the cultures used (Table 1).

Microbial enumeration of *Lactobacillus* species confirmed the fermentability of the peanut slurry samples. Both settings (fresh cheese culture alone, and fresh cheese culture added with *Propionibacterium freudenreichii* subsp. *globosum*) resulted in significant microbial growth for lactobacilli (Table 1).

Table 1. Fermentability of peanut slurry measured as pH and microbial enumeration of *Lactobacillus*. Mean values and standard deviation are reported.

Fermentability	UC	FC	UCP	FCP
pH	6.78 ± 0.03 [a]	4.74 ± 0.06 [b]	6.67 ± 0.04 [a]	4.70 ± 0.06 [b]
Microbial Enumeration (CFU/g)	6 × 10^3	TMC [1]	1 × 10^3	TMC [1]

UC = unfermented peanut slurry with the fresh cheese culture; FC = fermented peanut slurry with the fresh cheese culture; UCP = unfermented peanut slurry with the fresh cheese culture and *Propionibacterium freudenreichii* subsp. *Globosum;* (FCP) = fermented peanuts slurry with the fresh cheese culture *Propionibacterium freudenreichii* subsp. *globosum*. Different letters present a significant difference ($p < 0.05$) between the same sample, determined by one-way ANOVA and HSD test. [1] TMC = Too Many to Count.

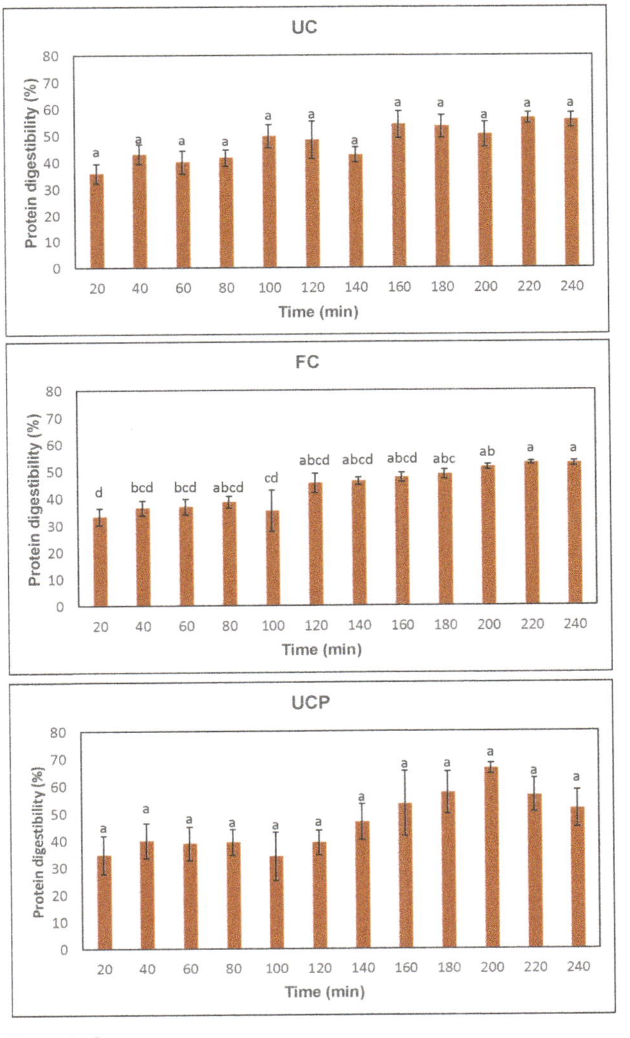

Figure 2. *Cont.*

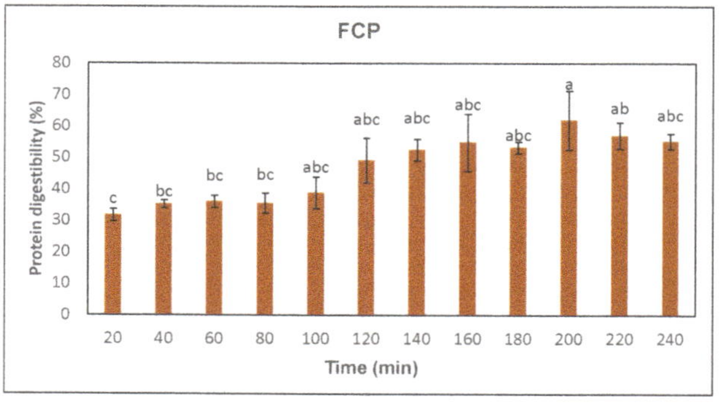

Figure 2. Changes in protein digestibility of peanut slurry. UC = unfermented peanut slurry with the fresh cheese culture; FC = fermented peanut slurry with the fresh cheese culture; UCP = unfermented peanut slurry with the fresh cheese culture and *Propionibacterium freudenreichii* subsp. *Globosum;* FCP = fermented peanuts slurry with the fresh cheese culture *Propionibacterium freudenreichii* subsp. *globosum*. Different letters present a significant difference (α = 0.05) between the same sample, determined by one-way ANOVA and HSD test.

3.2. Protein Digestibility

Figure 2 presents the results of the in vitro protein digestibility. The protein digestibility increased through the simulated gastric and intestinal digestions in all samples (Figure 2). The sample fermented with the combination of the fresh cheese culture and *Propionibacterium freudenreichii* subsp. *globosum* exhibited higher protein digestibility after 100 min than the sample fermented with the fresh cheese culture (Figure 2). Significant differences were observed in both the sample fermented with the fresh cheese culture after 180 min and the sample fermented with cheese culture and *Propionibacterium freudenreichii* subsp. *globosum* after 200 min (Figure 2). No significant difference was observed among different samples at the same digestion time points (Figure 2).

3.3. Total Amino Acid Profile

Quantification of total amino acids revealed abundance of glutamic acid (Glu), asparagine (Asn). As expected for legumes, lysine (Lys) was more abundant than methionine (Met). Fermentation with either *Lactococcus* alone or in combination with *Propionibacterium freudenreichii* subsp. *globosum* did not significantly alter the amino acid profile of peanuts (Figure 3).

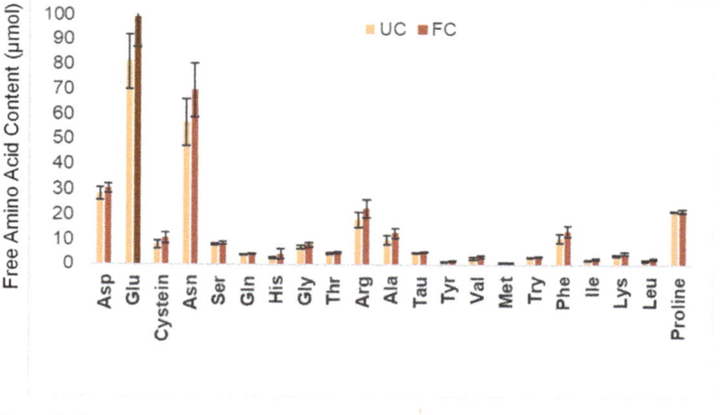

Figure 3. *Cont.*

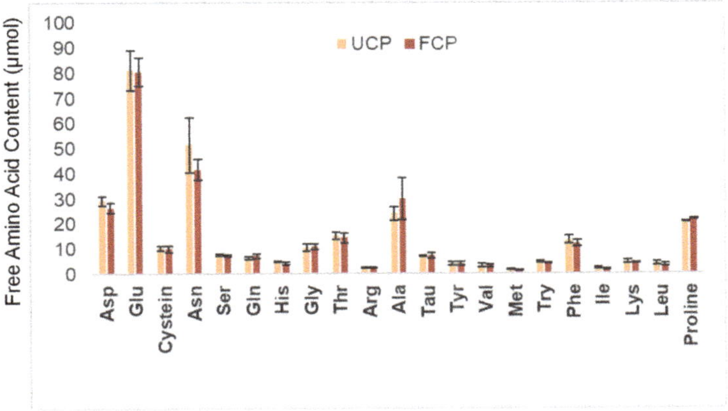

Figure 3. Changes in the total amino acid content. UC = unfermented peanut slurry with the fresh cheese culture; FC = fermented peanut slurry with the fresh cheese culture; UCP = unfermented peanut slurry with the fresh cheese culture and *Propionibacterium freudenreichii* subsp. *Globosum;* FCP = fermented peanuts slurry with the fresh cheese culture *Propionibacterium freudenreichii* subsp. *globosum*. Results are expressed as μmol per dry matter. Statistically significant difference was determined by one-way ANOVA with HSD test (α = 0.05).

3.4. Free Amino Acid Profile

Figure 4 shows the changes in the free amino acid profile before and after fermentation. The levels of all free amino acids increased after fermentation with the fresh cheese culture (Figure 4). On the other hand, fermentation with the combination of the fresh cheese culture and *Propionibacterium freudenreichii* subsp. *globosum* decreased the amount of all amino acids except for Gln, Ala, and Pro (Figure 4).

3.5. Texture

Textural analysis of peanut slurry revealed changes induced by fermentation. Specifically, the cheese culture addition resulted in a drastically softer texture: 52.8 vs. 220 g (Table 2). Softening of the fermented peanut slurry represented a four-fold reduction in hardness. On the contrary, *Propionibacterium freudenreichii* subsp. *globosum* addition did not produce the same results. This result correlated with the changes observed in protein digestibility (Figure 2).

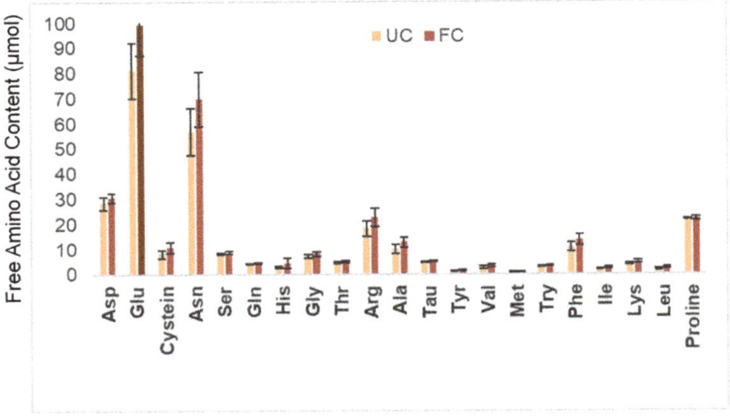

Figure 4. *Cont.*

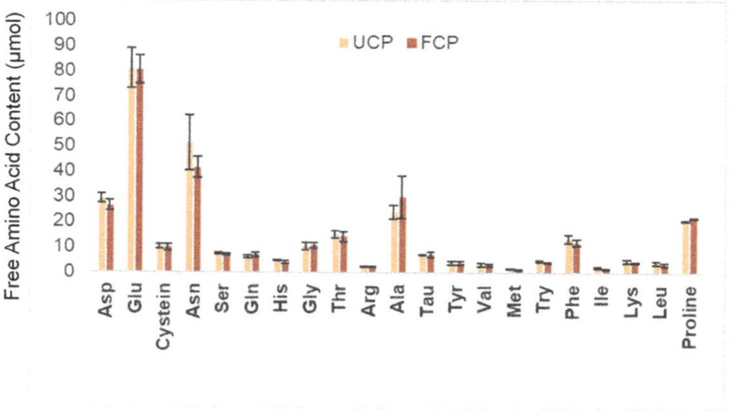

Figure 4. Changes in the free amino acid content. UC = unfermented peanut slurry with the fresh cheese culture; FC = fermented peanut slurry with the fresh cheese culture; UCP = unfermented peanut slurry with the fresh cheese culture and *Propionibacterium freudenreichii* subsp. *Globosum;* FCP = fermented peanuts slurry with the fresh cheese culture *Propionibacterium freudenreichii* subsp. *globosum.* Results are expressed as μmol per dry matter. Statistically significant difference was determined by one-way ANOVA with HSD test ($\alpha = 0.05$).

Table 2. Texture profile analysis of the slurry, both unfermented and fermented. Mean values and standard deviation are reported.

Texture Profile	UC	FC	UCP	FCP
Hardness (g)	220 ± 108 [a]	52.8 ± 4.8 [b]	141 ± 47 [ab]	100 ± 3 [ab]
Adhesiveness (g·mm)	82.9 ± 19.7 [b]	88.5 ± 16 [b]	299 ±175 [a]	173 ± 3 [ab]

UC = unfermented peanut slurry with the fresh cheese culture; FC = fermented peanut slurry with the fresh cheese culture; UCP = unfermented peanut slurry with the fresh cheese culture and *Propionibacterium freudenreichii* subsp. *Globosum;* (FCP) = fermented peanuts slurry with the fresh cheese culture *Propionibacterium freudenreichii* subsp. *globosum.* Different letters present a significant difference ($p < 0.05$) between the same sample, determined by one-way ANOVA and HSD test.

No significant differences were observed in terms of adhesiveness, yet larger variation was observed in the *Propionibacterium freudenreichii* subsp. *globosum*-containing recipe (Table 2).

4. Discussion

4.1. Fermentability

There was a significant decrease in the pH of the raw peanut slurry after fermentation (Table 1). A possible explanation for this result could be that acids produced by microbes during fermentation resulted in the pH decrease in the samples. Both *Lactococcus lactis* and *Leuconostoc*, which belong to lactic acid bacteria, ferment sugars to produce lactic acid, subsequently lowering the pH [15]. *Propionibacterium* also ferments sugars, producing propionic acid that reduces the pH [16]. Additionally, it was reported that *Propionibacterium freudenreichii* subsp. *globosum* consumed lactate and produced both propionate and acetate within 46 h when incubated at 30 °C [17]. Therefore, the decrease in the pH of the samples can be attributed to lactic acid, propionic acid, and acetate produced by the microbes during fermentation. The sugar profile of peanuts is mostly comprised of a disaccharide (sucrose) and oligosaccharides (stachyose and raffinose) [18]. Whereas sucrose can be fermented by numerous microorganisms, oligosaccharides are efficiently fermented by probiotic microorganisms such as *Lactococcus*. Fermentation of sugars by *Lactococcus lactis* has been shown to produce lactic acid, 2,3-butanediol, and ornithine (an amino acid) [19].

2,3-butanediol is a known plasticizer used as precursor in the rubber industry [20]. In addition, it delivers a creamy and fruity flavor [21]. The presence of this diol might explain the beneficial effects on protein solubility and texture.

Microbial growth was observed in both samples (Table 1). Peanuts can be an excellent fermentation matrix for probiotic bacteria such as *Lactobacillus* and *Propionibacterium* species. This is due to peanuts' nutritional value, delivering as much as 24 g of protein per 100 g sample, as well as 4.9 g of sugar [22]. In addition, peanuts contain nutritionally relevant levels of micronutrients such as vitamins B1, B2, B3, B6, B9, and choline [22] and minerals such as calcium, iron, magnesium, phosphorous, potassium, sodium, zinc, copper, manganese, and selenium [23]. These nutrients support microbial growth. Production of a slurry involves grinding peanuts into fine particles. This process can release amino acids, thus providing microorganisms with easily digestible nitrogen [24]. It must be stated that the initial number of probiotics was lower than in previous studies: approximately 3 vs. 6–8 CFU/g [24,25] due to the supplier recommendation for the cultures used. Post-fermentation results were indicated as TMC, as opposed to specific values. The goal of this study was not to quantify the probiotic level, but rather to verify whether fermentation of this matrix is possible.

4.2. Protein Digestibility

All samples demonstrated an increase in protein digestibility through the in vitro digestion process (Figure 2). The same trend was observed in a previous study. Ketnawa and Ogawa [8] fermented soybeans with *Bacillus natto* at 40 °C for 18 h. The protein digestibility of soaked, boiled, and fermented soybeans increased significantly as the simulated digestion time increased [8]. Despite the increase in protein digestibility in all samples, there were no significant differences observed in the samples before fermentation. One possible explanation for this might be the non-uniformity of the raw peanut slurry due to the amino acid profile. Raw peanuts mainly consist of both water-soluble and insoluble amino acids. In a previous study, it was revealed that raw peanuts mainly consisted of glutamic acid (859 ± 23 mg/kg dry weight), phenylalanine (555 ± 84 mg/kg dry weight), asparagine (238 ± 79 mg/kg dry weight), and valine (208 ± 9 mg/kg dry weight) [5]. It can be inferred that the insoluble proteins contribute to the non-uniformity of the peanut slurry before fermentation. This study could be improved by agitation before sampling to obtain uniform products.

Significant increases in protein digestibility during intestinal digestion were observed in the raw peanut slurry fermented with the fresh cheese culture after 180 min (Figure 2). The increases in protein digestibility can be attributed to small peptides produced by *Lactococcus lactis* and *Leuconostoc* in the fresh cheese culture, due to the production of proteases. *Lactococcus lactis* are proteolytic microbes that produce proteases during cell growth [25,26]. In a previous study, *Lactococcus lactis* was used as a proteolytic microbe to obtain bioactive peptides from soymilk through fermentation [27]. After fermentation, the soymilk had 2875 U/mL of protease activity, and the degree of hydrolysis value reached 44.32%. The degree of hydrolysis was assessed with the trinitro-benzene-sulfonic acid assay and calculated as the ratio between the increase in amino acid content after hydrolysis and the total peptide content, with the result expressed as a percentage [27]. *Leuconostoc* may also have a proteolytic ability. Rizzello and collaborators [28] fermented 300 g of a fava bean flour and water mixture using 6 log cfu/g of *Leuconostoc kimchi* at 25 °C for 48 h. After the fermentation, the peptide content significantly increased from 16.03 ± 0.94 g/kg to 20.18 ± 1.12 g/kg [28]. Researchers concluded that proteases produced by *Leuconostoc kimchi* hydrolyzed proteins, and the hydrolyzed peptide fractions increased the protein digestibility significantly, which increased from around 55% to 65% during the simulated gastrointestinal digestion. Further studies may explore the optimal temperature to improve the protein digestibility using the fresh cheese culture. Both *Lactococcus lactis* and *Leuconostoc* are mesophilic bacterium. While the optimal growth temperature of *Lactococcus lactis* was around 30 °C, the optimal temperature of *Leuconostoc* for maximum growth was from 34 to 36 °C [29,30]. In this study, the raw peanut slurry was incubated at 38 °C which was higher than the optimal one suggested by previous studies. Further research could assess the optimal temperature to maximize the protein digestibility of the fermented raw peanut slurry using the fresh cheese culture. The moisture content of the slurry was 70% for all samples treated, providing sufficient moisture for the proteolysis to take place.

While the raw peanut slurry fermented with a combination of the fresh cheese culture and *Propionibacterium freudenreichii* subsp. *globosum* demonstrated higher protein digestibility after 100 min than the sample fermented solely with the fresh cheese culture, significant differences were only observed at the 200 min (Figure 2). One potential explanation for this could be the formation of aggregates in the samples containing *Propionibacterium freudenreichii* subsp. *globosum*, contributing to the non-homogeneity of the peanut peptides. The proteinase activity of *Propionibacterium freudenreichii* subsp. *globosum* may be attributed to the following enzymes: aminopeptidase, iminopeptidase, X-prolyl dipeptidyl aminopeptidase, endopeptidase, two different oligopeptidases, and carboxypeptidase [31].

In a previous study, it was found that *Propionibacterium freudenreichii* subsp. *globosum* produced extracellular polysaccharides in a yeast extract–lactate medium and formed aggregates on a yeast extract–lactate plate [32]. The researchers concluded that the produced extracellular polysaccharides contributed to the aggregate characteristics of the *Propionibacterium freudenreichii* subsp. *globosum*. Based on this, it could be inferred that the aggregates formed by *Propionibacterium freudenreichii* subsp. *globosum* producing polysaccharides contributed to the non-homogeneity of the fermented peanut slurry. Further research could examine the protein digestibility of fermented raw peanut slurry using the combination of the fresh cheese culture and *Propionibacterium freudenreichii* subsp. *globosum* after agitation.

4.3. Total Amino Acid Profile

The total amount of acids in the slurries ranged from 40.4 to 43.7 µmol, without significant changes due to fermentation. This agreed with the previous literature on peanuts: total amino acids 45 µmol [33]. The most abundant amino acids found in the peanut slurry were Glu and Asp. This result also agreed with the previous literature [32]. Within essential amino acids, lysine (Lys) was three times more abundant than methionine (Met), as expected for legume seeds [30].

4.4. Amino Acid Profile

All free amino acid concentrations increased after fermentation using the fresh cheese culture (Figure 4). The increase can be attributed to proteolytic microbes in the fresh cheese culture hydrolyzing proteins in the raw peanut slurry. In a previous study, soybeans were fermented using *Bacillus natto* at 40 °C for 18 h [8]. After the fermentation, all free amino acid concentrations increased except for Arg [8]. The researchers concluded that hydrolyzation during fermentation with the proteolytic microbe improved the free amino acid profile. Although it can be inferred that fermentation improved the free amino acid profile of raw peanut slurry, this research was unable to determine significant differences. Further research could identify the specific amino acids that experience significant increases during fermentation with the fresh cheese culture. Further studies could also examine the changes in free amino acid profile after in vitro digestion to estimate nutritional availability in the body.

On the other hand, fermentation with the combination of the fresh cheese culture and *Propionibacterium freudenreichii* subsp. *globosum* led to a decrease in the concentration of all amino acids, except for Gln, Ala, and Pro. It could be inferred that *Propionibacterium freudenreichii* subsp. *globosum* consumed the free amino acids produced by the fresh cheese culture during fermentation. In a previous study, *Propionibacterium freudenreichii globosum* was inoculated in yeast extract–lactate medium and incubated at 30 °C [34]. The researchers found that the *Propionibacterium freudenreichii* subsp. *globosum* strain consumed all free amino acids except for Thr, Gln, Val, Met, Lie, Try, His, and Pro in yeast extract–lactate medium after 3 days of incubation [32]. Based on this, it could be inferred that *Propionibacterium freudenreichii* subsp. *globosum* might consume the amino acids hydrolyzed by the fresh cheese culture. Aburjaile and coauthors [34] also mentioned that the concentration of several free amino acids increased after 9 days of fermentation. Future studies could investigate the change in free amino acid concentrations over different fermentation periods.

Organoleptic quality is crucial. The current study estimated sensory quality instrumentally: free amino acids and peptides affect aroma and flavor, hardness and adhesiveness affect texture. Free amino acids contribute to the aroma and flavor of food. Fermentation with the cheese culture freed glutamic acid and asparagine (Figure 4). Glutamic acid is known to produce the umami taste [35], as some hydrolysate peanut peptides do [36]. This is a flavor profile that typically entices consumers, as it occurs with cheese, meat, and mushrooms. Interestingly, asparagine flavor profile depends on its structure. While L-asparagine is described as tasteless, D-asparagine was described as having an intense sweet flavor [37]. These changes in the amino acidic profile can potentially increase consumer acceptance of the fermented peanut slurry. In addition, *Propionibacterium freudenreichii* subsp. *globosum* is known to reduce the acidity of fermented food. Therefore, it is possible that fermentation of peanut slurry with *Lactobacillus lactis* might increase consumer acceptability of this highly digestible protein source. No significant changes in pH and fat content were observed among fermented and unfermented samples (Table 1).

4.5. Texture

Fermentation using the fresh cheese culture significantly decreased the hardness of the peanut slurry, resulting in a more homogeneous product with a smaller standard deviation (Table 2). The increased protein solubility (Figure 2), which is the second most abundant component after water, likely explains the decrease in the hardness [38,39]. Another factor to consider is pH. Protein structure changes at different pH values. Peanut protein was determined to be more soluble, thus leading to a softer texture, at pH 7 than 5 [40]. Therefore, increased solubility of the acidified slurry (post-fermentation) is remarkable and likely attributed to protein hydrolysis. One possible explanation is that lactic fermentation of legume protein increases its emulsifying properties [41]. The peanut slurry was abundant in oil and water, resulting in an unstable matrix, as shown by the high standard deviation of the raw samples (Table 2). Higher emulsifying ability results in a homogeneous structure and lower surface tension, hence, a softer texture. *Propionibacterium freudenreichii* subsp. *globosum* metabolizes lactic acid into propionic acid, pyruvic acid, carbon dioxide, and vitamin B12 [42]. The main reason for using this microorganism was the synthesis of vitamin B12 in plant-based food.

On the other hand, no significant difference was observed in the sample fermented using both the fresh cheese culture and *Propionibacterium freudenreichii* subsp. *globosum* (Table 2). Only the unfermented sample with both cultures (FCP) was drastically more adhesive than the other three treatments. It has been hypothesized that the propionic culture contained compounds such as complex sugars which might contribute to the sticky texture. The lower standard deviation observed in the fermented slurry might also be attributed to the increased protein solubility. Peanuts are lipid-rich seeds and have low affinity to water, which are composed of approximately 50% oil and 25% protein [23]. Enhanced protein solubility may increase the water affinity of ground peanuts, potentially resulting in a more homogeneous peanut slurry.

On the contrary, adhesiveness was not affected by any treatment. This indicates that soluble protein did not interact with moisture distribution within the slurry. Consequently, it can be stated that the peanut slurry fermented with cheese culture was significantly softer.

5. Conclusions

This study has shown that protein digestibility of peanuts can be increased by means of lactic fermentation. Specifically, incubation of a peanut slurry with *Lactobacillus* species and *Leuconostoc* enhanced protein digestibility while decreasing slurry hardness. The slurry fermented with the fresh cheese culture exhibited an increase in protein solubility, from 35 to 50%, after a 240-h in vitro digestion. Simultaneously, slurry hardness decreased from 220 to 53 g upon lactic fermentation, and from 141 to 40 g with *Propionibacterium freudenreichii* subsp. *globosum*.

Probiotic bacteria grew abundantly in this highly nutritious media, releasing amino acids that contribute to umami and sweet taste. Fermentation with *Propionibacterium freudenreichii* subsp. *globosum* did not significantly increase the protein solubility, while it did soften the texture. This study highlighted the benefits of fermenting peanuts with these starter cultures: more digestible protein, sweeter umami taste, and softer texture. Furthermore, *P. globosum* has the potential to synthesize vitamin B12, an essential nutrient lacking in plant-based foods. Potential challenges include fermentation time and cost.

Author Contributions: Conceptualization, L.S. (Luca Serventi) and L.S. (Letitia Stipkovits); methodology, L.S. (Luca Serventi), L.S. (Letitia Stipkovits), and Y.M.; software, L.S. (Luca Serventi) and A.S.; validation, L.S. (Luca Serventi) and A.S.; formal analysis, A.S., L.S. (Letitia Stipkovits) and Y.M.; investigation, A.S., L.S. (Letitia Stipkovits) and Y.M.; resources, L.S. (Luca Serventi).; data curation, A.S. and L.S. (Luca Serventi); writing—original draft preparation, A.S. and L.S. (Luca Serventi); writing—review and editing, L.S. (Luca Serventi), L.S. (Letitia Stipkovits) and Y.M.; visualization, L.S. (Luca Serventi); supervision, L.S. (Luca Serventi); project administration, L.S. (Luca Serventi). All authors have read and agreed to the published version of the manuscript.

Funding: This research received no external funding.

Data Availability Statement: The data used to support the findings of this study can be made available by the corresponding author upon request.

Acknowledgments: The authors would like to thank Jenny Zhao (Lincoln University) for performing the analyses of total and free amino acids. In addition, they would like to acknowledge Naiyi Chen, Ziyu Liao, Yue Hu and Rongzheng Yu (Lincoln University and Zhongkai University) for their contribution to the preliminary research, data collection and image acquisition/creation (graphical abstract).

Conflicts of Interest: The authors declare no conflict of interest.

References

1. Polak, R.; Phillips, E.M.; Campbell, A. Legumes: Health Benefits and Culinary Approaches to Increase Intake. *Clin. Diabetes* **2015**, *33*, 198–205. [CrossRef]
2. USDA. 2018. Available online: https://fdc.nal.usda.gov/fdc-app.html#/food-details/172430/nutrients (accessed on 18 July 2023).
3. Sá, A.G.A.; Moreno, Y.M.F.; Carciofi, B.A.M. Food processing for the improvement of plant proteins digestibility. *Crit. Rev. Food Sci. Nutr.* **2020**, *60*, 3367–3386. [CrossRef]
4. The Ministry for Primary Industries; Plant and Food Research. The Evolution of Plant Protein. 2018. Available online: https://www.mpi.govt.nz/dmsdocument/29141-The-Evolution-of-Plant-Protein-Assessing-Consumer-Response-Report (accessed on 18 July 2023).
5. Miyagi, A.; Ogaki, Y. Chemical processes in peanut under thermal treatment. *J. Food Meas. Charact.* **2014**, *8*, 305–315. [CrossRef]
6. Food and Agriculture Organization of the United Nations. Dietary Protein Quality Evaluation in Human Nutrition. 2013. Available online: https://www.fao.org/ag/humannutrition/35978-02317b979a686a57aa4593304ffc17f06.pdf (accessed on 18 July 2023).
7. Chen, C.C.; Shih, Y.C.; Chiou, P.W.S.; Yu, B. Evaluating Nutritional Quality of Single Stage- and Two Stage-fermented Soybean Meal. *Asian-Australas. J. Anim. Sci.* **2010**, *23*, 598–606. [CrossRef]
8. Ketnawa, S.; Ogawa, Y. In vitro protein digestibility and biochemical characteristics of soaked, boiled and fermented soybeans. *Sci. Rep.* **2021**, *11*, 14257. [CrossRef]
9. Griffith, L.; Castell-Perez, M.; Griffith, M. Effects of blend and processing method on the nutritional quality of weaning foods made from select cereals and legumes. *Cereal Chem.* **1998**, *75*, 105–112. [CrossRef]
10. Mugula, J.; Lyimo, M. Evaluation of the nutritional quality and acceptability of sorghum-based tempe as potential weaning foods in Tanzania. *Int. J. Food Sci. Nutr.* **2000**, *51*, 269–277.
11. Taye, Y.; Degu, T.; Fesseha, H.; Mathewos, M. Isolation and identification of lactic acid bacteria from cow milk and milk products. *Sci. World J.* **2021**, *2021*, 4697445. [CrossRef]
12. Uraipong, C.; Zhao, J. In vitro digestion of rice bran proteins produces peptides with potent inhibitory effects on -glucosidase and angiotensin I converting enzyme. *J. Sci. Food Agric.* **2018**, *98*, 758–766. [CrossRef]

13. Almeida, C.C.; Monteiro, M.L.G.; da Costa-Lima, B.R.C.; Alvares, T.S.; Conte-Junior, C.A. In vitro digestibility of commercial whey protein supplements. *LWT-Food Sci. Technol.* **2015**, *61*, 7–11. [CrossRef]
14. Chen, W.; Chiu, H.T.; Feng, Z.; Maes, E.; Serventi, L. Effect of spray-drying and freeze-drying on the composition, physical properties, and sensory quality of pea processing water (Liluva). *Foods* **2021**, *10*, 1401. [CrossRef]
15. Makarova, K.; Slesarev, A.; Wolf, Y.; Sorokin, A.; Mirkin, B.; Koonin, E.; Pavlov, A.; Pavlova, N.; Karamychev, V.; Polouchine, N.; et al. Comparative genomics of the lactic acid bacteria. *Proc. Natl. Acad. Sci. USA* **2006**, *103*, 15611–15616. [CrossRef]
16. Hsu, S.T.; Yang, S.T. Propionic-acid fermentation of lactose by *propionibacterium-acidipropionici*—Effects of pH. *Biotechnol. Bioeng.* **1991**, *38*, 571–578. [CrossRef]
17. Dank, A.; Mastrigt, O.V.; Boeren, S.; Lillevang, S.K.; Abee, T.; Smid, E.J. *Propionibacterium freudenreichii* thrives in microaerobic conditions by complete oxidation of lactate to CO_2. *Environ. Microbiol.* **2021**, *23*, 3116–3129. [CrossRef]
18. Basha, S.M. Soluble sugar composition of peanut seed. *J. Agric. Food Chem.* **1992**, *40*, 780–783. [CrossRef]
19. Golomb, B.L.; Marco, M.L. Lactococcus lactis metabolism and gene expression during growth on plant tissues. *J. Bacteriol.* **2015**, *197*, 371–381. [CrossRef]
20. Narisetty, V.; Zhang, L.; Zhang, J.; Lin, C.S.K.; Tong, Y.W.; Show, P.L.; Bhatia, S.K.; Misra, A.; Kumar, V. Fermentative production of 2, 3-Butanediol using bread waste–A green approach for sustainable management of food waste. *Bioresour. Technol.* **2022**, *358*, 127381. [CrossRef]
21. De Faveri, D.; Torre, P.; Molinari, F.; Perego, P.; Converti, A. Carbon material balances and bioenergetics of 2, 3-butanediol bio-oxidation by Acetobacter hansenii. *Enzyme Microb. Technol.* **2003**, *33*, 708–719. [CrossRef]
22. Bonku, R.; Yu, J. Health aspects of peanuts as an outcome of its chemical composition. *Food Sci. Hum. Wellness* **2020**, *9*, 21–30. [CrossRef]
23. Settaluri, V.S.; Kandala, C.V.K.; Puppala, N.; Sundaram, J. Peanuts and their nutritional aspects—Review. *Food Nutr. Sci.* **2012**, *3*, 25267. [CrossRef]
24. Singh, B.; Singh, U. Peanut as a source of protein for human foods. *Plant Foods Hum. Nutr.* **1991**, *41*, 165–177. [CrossRef]
25. Chavan, M.; Gat, Y.; Harmalkar, M.; Waghmare, R. Development of non-dairy fermented probiotic drink based on germinated and ungerminated cereals and legume. *LWT* **2018**, *91*, 339–344. [CrossRef]
26. Wang, S.; Chelikani, V.; Serventi, L. Evaluation of chickpea as alternative to soy in plant-based beverages, fresh and fermented. *LWT* **2018**, *97*, 570–572. [CrossRef]
27. Anggraeni, S.L.; Jayus, J.; Ratnadewi, A.; Nurhayati, N. Edamame protein hydrolysis using *Lactococcus lactis*, *Lactobacillus bulgaricus* and *Lactobacillus paracasei* produce short peptides with higher antioxidant potential. *Biodiversitas* **2022**, *23*, 7. [CrossRef]
28. Rizzello, C.; Coda, R.; Wang, Y.; Verni, M.; Kajala, I.; Katina, K.; Laitila, A. Characterization of indigenous *Pediococcus pentosaceus, Leuconostoc kimchii, Weissella cibaria* and *Weissella confusa* for faba bean bioprocessing. *Int. J. Food Microbiol.* **2019**, *302*, 24–34. [CrossRef]
29. Cooper, R.K.; Collins, E.B. Influences of Temperature on Growth of *Leuconostoc cremoris*. *J. Dairy Sci.* **1978**, *61*, 1085–1088. [CrossRef]
30. Enan, G.; Abdel-Shafi, S.; Ouda, S.; Negm, S. Novel antibacterial activity of *lactococcus lactis subspecies lactis z11* isolated from zabady. *J. Biomed. Sci.* **2013**, *9*, 174–180.
31. Rattray, F.P.; Eppert, I. Secondary cultures. In *Encyclopedia of Dairy Sciences*, 2nd ed.; Elsevier: Amsterdam, The Netherlands, 2022; pp. 567–573. [CrossRef]
32. Riedel, K.H.J.; Britz, T.J. *Propionibacterium* species-diversity IN anaerobic digesters. *Biodivers. Conserv.* **1993**, *2*, 400–411. [CrossRef]
33. Liu, K.; Liu, Y.; Chen, F. Effect of storage temperature on lipid oxidation and changes in nutrient contents in peanuts. *Food Sci. Nutr.* **2019**, *7*, 2280–2290. [CrossRef]
34. Aburjaile, F.F.; Rohmer, M.; Parrinello, H.; Maillard, M.B.; Beaucher, E.; Henry, G.; Nicolas, A.; Madec, M.N.; Thierry, A.; Parayre, S.; et al. Adaptation of *Propionibacterium freudenreichii* to long-term survival under gradual nutritional shortage. *BMC Genom.* **2016**, *17*, 1007. [CrossRef]
35. Ninomiya, K. Science of umami taste: Adaptation to gastronomic culture. *Flavour* **2015**, *4*, 13. [CrossRef]
36. Zhang, J.; Zhao, M.; Su, G.; Lin, L. Identification and taste characteristics of novel umami and umami-enhancing peptides separated from peanut protein isolate hydrolysate by consecutive chromatography and UPLC–ESI–QTOF–MS/MS. *Food Chem.* **2019**, *278*, 674–682. [CrossRef]
37. Gal, J. The discovery of stereoselectivity at biological receptors: Arnaldo Piutti and the taste of the asparagine enantiomers—History and analysis on the 125th anniversary. *Chirality* **2012**, *24*, 959–976. [CrossRef]
38. Sun, Q.; Xiong, C.S.L. Functional and pasting properties of pea starch and peanut protein isolate blends. *Carbohydr. Polym.* **2014**, *101*, 1134–1139. [CrossRef]
39. Wu, H.; Wang, Q.; Ma, T.; Ren, J. Comparative studies on the functional properties of various protein concentrate preparations of peanut protein. *Food Res. Int.* **2009**, *42*, 343–348. [CrossRef]
40. Emkani, M.; Oliete, B.; Saurel, R. Effect of lactic acid fermentation on legume protein properties, a review. *Fermentation* **2022**, *8*, 244. [CrossRef]

41. Ge, J.; Sun, C.X.; Sun, M.; Zhang, Y.; Fang, Y. Introducing panda bean (*Vigna umbellata (Thunb.) Ohwi et Ohashi*) protein isolate as an alternative source of legume protein: Physicochemical, functional and nutritional characteristics. *Food Chem.* **2022**, *388*, 133016. [CrossRef]
42. Dank, A.; Biel, G.; Abee, T.; Smid, E.J. Microaerobic metabolism of lactate and propionate enhances vitamin B12 production in *Propionibacterium freudenreichii*. *Microb. Cell Factories* **2022**, *21*, 225. [CrossRef]

foods

Article

Protective Effect of Long-Term Fermented Soybeans with Abundant *Bacillus subtilis* on Glucose and Bone Metabolism and Memory Function in Ovariectomized Rats: Modulation of the Gut Microbiota

Hee-Jong Yang [1], Ting Zhang [2], Yu Yue [3], Su-Ji Jeong [1], Myeong-Seon Ryu [1], Xuangao Wu [2], Chen Li [3], Do-Yeon Jeong [1,*] and Sunmin Park [2,3,*]

1 Department of R & D, Microbial Institute for Fermentation Industry, Sunchang-gun 56048, Republic of Korea; godfiltss@naver.com (H.-J.Y.); yo217@naver.com (S.-J.J.); rms6223@naver.com (M.-S.R.)
2 Department of Bioconvergence, Hoseo University, Asan-si 31499, Republic of Korea; zhangting92925@gmail.com (T.Z.); niyani0@naver.com (X.W.)
3 Department of Food and Nutrition, Obesity/Diabetes Research Center, Hoseo University, Asan-si 31499, Republic of Korea; yuyue6491@gmail.com (Y.Y.); lic77732@gmail.com (C.L.)
* Correspondence: jdy2534@korea.kr (D.-Y.J.); smpark@hoseo.edu (S.P.); Tel.: +82-63-650-2001 (D.-Y.J.); +82-41-540-5345 (S.P.); Fax: +82-63-650-9590 (D.-Y.J.); +82-41-548-0670 (S.P.)

Abstract: We investigated the effects of different types of long-term fermented soybeans (traditionally made doenjang; TMD) on glucose and bone metabolism and memory function in ovariectomized (OVX) rats. The rats were categorized into six groups: Control, cooked unfermented soybeans (CSB), and four TMDs based on *Bacillus subtilis* (*B. subtilis*) and biogenic amine contents analyzed previously: high *B. subtilis* (HS) and high biogenic amines (HA; HSHA), low *B. subtilis* (LS) and HA (LSHA), HS and low biogenic amines (LA; HSLA), and LS and LA (LSLA). The rats in the CSB and TMD groups fed orally had a 4% high-fat diet for 12 weeks. Rats in the Control (OVX rats) and Normal-control (Sham-operated rats) groups did not consume CSB or TMD, although macronutrient contents were the same in all groups. Uterine weight and serum 17β-estradiol concentrations were much lower in the Control than the Normal-control group, but CSB and TMD intake did not alter them regardless of *B. subtilis* and biogenic amine contents. HOMA-IR, a measure of insulin resistance, decreased with TMD with high *B. subtilis* (HSLA and HSHA) compared to the Control group. In OGTT and IPGTT, serum glucose concentrations at each time point were higher in the Control than in the Normal-control, and HSLA and HSHA lowered them. Memory function was preserved with HSHA and HSLA administration. Bone mineral density decline measured by DEXA analysis was prevented in the HSHA and HSLA groups. Bone metabolism changes were associated with decreased osteoclastic activity, parathyroid hormone levels, and osteoclastic activity-related parameters. Micro-CT results demonstrated that TMD, especially HSLA and HSHA, preserved bone structure in OVX rats. TMD also modulated the fecal bacterial community, increasing *Lactobacillus*, *Ligalactobacillus*, and *Bacillus*. In conclusion, through gut microbiota modulation, TMD, particularly with high *B. subtilis* content, acts as a synbiotic to benefit glucose, bone, and memory function in OVX rats. Further research is needed to make specific recommendations for *B. subtilis*-rich TMD for menopausal women.

Keywords: traditionally made doenjang; glucose metabolism; memory function; bone mineral density; estrogen deficiency

Citation: Yang, H.-J.; Zhang, T.; Yue, Y.; Jeong, S.-J.; Ryu, M.-S.; Wu, X.; Li, C.; Jeong, D.-Y.; Park, S. Protective Effect of Long-Term Fermented Soybeans with Abundant *Bacillus subtilis* on Glucose and Bone Metabolism and Memory Function in Ovariectomized Rats: Modulation of the Gut Microbiota. *Foods* **2023**, *12*, 2958. https://doi.org/10.3390/foods12152958

Academic Editors: Jennifer Ahn-Jarvis and Brittany A. Hazard

Received: 17 July 2023
Revised: 3 August 2023
Accepted: 3 August 2023
Published: 4 August 2023

1. Introduction

Menopause occurs when the ovaries stop producing estrogen. Estrogen is responsible for several functions in the body, such as regulating the menstrual cycle, controlling energy, glucose, and lipid levels, maintaining bone mineral density (BMD), and cognitive function [1,2]. Menopausal symptoms include hot flashes, mood swings, night sweats, and insomnia. In addition, menopause raises the risk of metabolic diseases along with obesity [1,2]. The incidence and severity of menopausal symptoms may vary depending on the geographic region and cultural factors. Asian women have a lower incidence of menopausal symptoms than Western women, which may be potentially related to differences in dietary patterns, such as a high intake of soybeans.

Estrogen promotes the production and upregulation of glucose transporters, especially glucose transporter-4 (GLUT-4), in cell membranes through estrogen receptors (ESR)-1 and ESR-2, allowing efficient glucose uptake from the bloodstream [3,4]. Estrogen deficiency, conversely, can induce insulin resistance and weight gain while also being a well-known factor contributing to BMD loss [5]. While higher body mass index, particularly in the overweight and obese range, is generally associated with increased mechanical loading on bones and stimulating bone formation, postmenopausal women often experience lower cortical bone mass and strength despite elevated insulin resistance [5–7]. Insulin resistance in prediabetic and diabetic patients is also associated with cognitive dysfunction, potentially through increasing Tau phosphorylation [8]. Interestingly, estrogen deficiency affects bone health and has implications for cognitive function [8]. Estrogen plays a role in brain connectivity, and its deficiency can contribute to cognitive impairment and amyloid-beta deposition, leading to Alzheimer's disease development [9]. Low estrogen infusion to the brain enhances cognitive function in estrogen-deficient animal models [10]. Therefore, postmenopausal women with abdominal abnormal glucose metabolism are at a higher risk of experiencing osteoporosis and cognitive dysfunction.

Hormone replacement therapy with estrogen with or without progestogen improves menopausal symptoms. However, it may raise the risk of women-related cancers such as breast, ovarian, and uterine cancers. Alternative therapies involve natural remedies, dietary and lifestyle changes, herbal supplements, and other complementary medicines [11]. Phytoestrogens, mainly isoflavones, may protect against trabecular bone loss in postmenopausal women by enhancing bone formation and suppressing bone resorption [12]. Soybeans contain isoflavonoids known as phytoestrogens, and their intake is reported to improve bone density and skeletal muscle mass and reduce body weight after menopause [13]. However, fermented soybeans have isoflavone aglycones, which have greater bioavailability and may have better bone formation activity, and may improve cognitive function as observed in estrogen-deficient animal models [14–16]. Furthermore, some gut microbiota can convert daidzein in fermented soybeans, such as chungkookjang, into equol, the most potent phytoestrogen [17].

Traditionally made doenjang (TMD) is made by fermenting soybeans for about one year and is influenced by environmental factors, such as bacteria composition, fermentation temperature, seasonal variation, fermentation time, and salt contents [18]. TMD contains beneficial bacteria like *Bacillus*, *Lactobacillus*, *Pediococcus*, and *Weissella*, contributing to its potential as a synbiotic food [18]. However, some TMD products may also harbor low levels of harmful bacteria and contain biogenic amines during fermentation [18]. Biogenic amines are formed through the microbial decarboxylation of amino acids during fermentation. However, some *Bacillus subtilis* (*B. subtilis*) strains also have degradation activity of biogenic amines since different strains of bacteria have particular functions in fermented foods [19]. While TMD generally contains biogenic amines within safe levels (<200–500 mg/kg for fish and its products in CODEX) [20], it is essential to consider the varying quantities of these compounds in different TMD products. Excessive consumption of certain biogenic amines has been associated with adverse effects such as headaches, migraines, skin rashes, digestive issues, and allergic-like reactions [19]. However, the specific health implications of various biogenic amine quantities have not been extensively studied.

TMD intake improves the bioavailability of bioactive components in soybeans and menopausal symptoms in estrogen-deficient animal models [16]. It has also shown antiobesity activity in randomized clinical trials [21]. However, few studies have examined the TMD intake effect on glucose and bone metabolism and cognitive function in estrogen-deficient animal models. The present study aimed to investigate whether the TMD intake containing different quantities of *B. subtilis* and biogenic amines could improve glucose and bone metabolism and suppress cognitive impairment in OVX rats, known to be estrogen-deficient animal models.

2. Materials and Methods

2.1. Collection of TMD Products

Over 50 TMD products were purchased from different regions in Korea. Their characteristics were different due to varying annual temperatures, air pollution, and salt content. They were generally made by fermenting boiled soybeans with 10–13% salt for about 12 months outdoors [18,19]. Bacteria in TMD mainly originated from the air and were modulated by the salt content and bacterial community during fermentation [18,19]. Their compositions were analyzed for bacteria, biogenic amines, isoflavonoids, and sodium. The bacteria compositions in TMD were analyzed using the next-generation sequencing (NGS) in the 2.10 section based on previous studies [18,22].

The sodium content in TMD was quantified using ICP-AES (Thermo IRIS Intrepid II XDL, USA) following Korean MFDS guidelines. Protein digestion with nitric acid was performed prior to measurement. Methanol was mixed with TMD, and the filtrates were used to measure the sodium content. The oxygen and acetylene flows were set at 10.00 L/min and 2.50 L/min, respectively, with an air acetylene flame type, and the wavelength was set at 589.0 nm. Isoflavonoid glycone and aglycone contents in the filtrates were analyzed using HPLC (Agilent 1200 series, Agilent Technologies, Santa Clara, CA, USA) equipped with a Shiseido UG 120 column (4.6 × 250 mm, 5 μm, Osaka, Japan) according to the established procedure. As previously described, biogenic amine contents, including histamine and tyramine, were determined using HPLC analysis with a Cepcell Pak C18 column [23]. TMD was mixed with an internal standard and derivatized with sodium carbonate and dansyl chloride prior to measuring the biogenic amine contents. From the NGS results and biogenic amine contents of 50 TMD products, four TMD products were selected such as high and low *B. subtilis* at the cutoff of 70% of total bacteria and high and low biogenic amines at the cutoff of 300 mg/kg TMD. The four selected TMDs contained (1) High *B. subtilis* (HS) plus low biogenic amines (LA)(HSLA), (2) HS plus high biogenic amines (HA)(HSHA), (3) Low *B. subtilis* (LS) plus LA (LSLA), and (4) Low LS plus HA (LSHA).

2.2. Ovariectomy Procedure

Sixty female Sprague-Dawley rats (8 weeks old, weighing 163 ± 12 g) were obtained from DBL Co. Ltd. (Yeumsung-Kun, Republic of Korea) and individually housed in stainless-steel cages with a 12-h light/dark cycle at a room temperature of 23 °C. Following a week of acclimatization at the animal facility of Hoseo University, the rats underwent ovariectomy (OVX) surgery. An animal study was performed under the ethical guidelines and regulations, approved by the Hoseo University Animal Care and Use Committee (Approval No. HSIACUC-22-03), and aligned with the principles outlined in the Guide for the Care and Use of Laboratory Animals by the National Institutes of Health (NIH) in the USA. Both ovaries of sixty rats were dissected with scissors after subcutaneously injecting a ketamine/xylazine mixture (100 and 10 mg/kg body weight). An additional ten rats underwent a sham operation for ovariectomy.

2.3. Diet Preparation

All the groups were given a baseline high-fat diet (HFD), which diminished bone mineral density and bone microstructure [6]. This diet was formulated to contain 43.4%, 17.1%, and 39.5% of energy (En%), fat, protein, and carbohydrates, respectively, and 5.9 g salt/kg based on the AIN-93 diet for small animals [24]. The diet was supplemented with either 4% (*w/w*) of lyophilized doenjang or 4% (*w/w*) of cooked soybeans. As doenjang and soybeans contain fats, proteins, and carbohydrates, the nutrient composition of these ingredients was subtracted from the corresponding components in the corresponding diet to achieve uniform carbohydrate, fat, and protein composition across all diets. This adjustment ensured that all diets had equivalent carbohydrate, fat, and protein composition, allowing for a fair comparison of the effects of the interventions. The assigned starch, casein, vitamins, minerals, and either TMD or cooked soybeans were combined and mixed. Subsequently, lard and soybean oil were added to the mixture, and the ingredients were sifted to remove lumps. All the diets in the TMD and control groups had equivalent amounts of carbohydrates, proteins, fats, and sodium.

2.4. Experimental Design

Sixty OVX rats were randomly allocated to each group and given either cooked soybeans or four different TMD products. Each group was labeled based on the type of diet administered: (1) Control (high-fat diet without TMD or cooked soybeans), (2) cooked soybeans (CSB), lyophilized TMD with (3) HSHA, (4) HSLA, (5) LSHA, and (6) LSLA. In addition, ten sham-operated rats were included as Normal controls. All the diets had an equivalent macronutrient composition. Throughout the twelve-week study period, all rats were provided unrestricted access to food and water, and their food intake and body weight were monitored and recorded weekly.

2.5. Glucose Metabolism Measurement

After the 11-week TMD intervention, the animals underwent an oral glucose tolerance test (OGTT) following an overnight fasting period. During the test, they were orally administered 2 g of glucose per kg of body weight. Tail blood samples were collected at 10-min intervals for up to 120 min to measure serum glucose concentrations. Additionally, serum insulin concentrations were assessed at 0, 20, 40, 90, and 120 min. The trapezoidal rule was applied to calculate the mean area under the curve (AUC) for both serum glucose and insulin concentrations. Three days after the OGTT, the animals underwent an intraperitoneal insulin tolerance test (IPITT) following a 6-h fast. Their serum glucose levels were measured at 15-min intervals for 90 min after receiving an intraperitoneal insulin injection at a dosage of 0.75 U/kg of body weight.

2.6. Memory Function Measurement

A passive avoidance test was conducted using a shuttle box apparatus equipped with two dark/light compartments [25]. A rat was initially placed in the light compartment. When it entered the dark chamber, electrostimulation (75 V, 0.2 mA, 50 Hz) was delivered in two acquisition trials with an 8-h interval. In the third trial, the latency taken to enter the dark chamber was assessed without electrostimulation for 600 s. A longer latency indicates better memory function.

For the forced swimming test, the rat was food-deprived for 8 h and then subjected to a 10-min pretest in a clear acrylic cylinder. The cylinder was 60 cm in height and 30 cm in width, and it was filled with 45 cm of water maintained at a temperature of 24 ± 1 °C. After 24 h, a 5-min forced swimming test was conducted in the same cylinder. During this test, the rat's movements were recorded and scored for mobile behaviors, such as swimming and climbing, as well as immobile behaviors [25]. The total time spent on active and passive behaviors was calculated. Rats showing longer active time during the forced swimming session were considered less depressed.

The novel object recognition test was performed by modifying previously established methods [25]. On the first two days, rats were given 5 min to explore a plexiglass box. During the training phase, the box was divided into four areas, and two identical objects were placed in the center of the diagonal area. Rats were then placed in the center of the box and allowed to explore the objects for 5 min freely. Afterward, the objects and instruments were cleaned with 75% ethanol to remove residual odor. On the final day of the experiment, one of the objects was replaced with a new object that was similar in size but differed in shape and color from the original object. The rats' exploration behavior, which included nose/paw contact (excluding sitting on or turning the object around), was recorded for both the old and new objects. The time spent exploring each object was measured. The recognition index was calculated by dividing the time spent exploring the new object by the total time spent exploring both objects. If the recognition index was lower than 50%, it was considered incidental, indicating poor recognition. Higher values of the recognition index indicated good recognition of the new object.

2.7. BMD Measurement by Dual-Energy X-ray Absorptiometry (DEXA) and Sample Collection

Two days after the IPITT, the body composition of the rats was assessed using DEXA while they were under anesthesia with a ketamine/xylazine mixture. The DEXA measurements were performed using a DEXA machine supplied by Norland Medical Systems Inc., Fort Atkinson, WI, USA. Before and after the experiments, a calibration phantom was used to ensure accurate measurements of the rats' body composition. The anesthetized rat was positioned in a prone position with its rear legs securely in external rotation, and the hip, knee, and ankle joints were flexed at 90° for the DEXA scan. Following the scan, the BMD was determined using the DEXA machine's software designed explicitly for rodent BMD analysis [26].

After completing the DEXA analysis, the rats were euthanized. While under anesthesia induced by a combination of ketamine and xylazine, blood samples were drawn from the inferior vena cava. The femur was then carefully dissected following the 12-week treatment period. Next, the epididymal and retroperitoneal fat masses were excised and weighed, along with the uteri. The uterus index was calculated by dividing the uterus weight by the body weight of the rats. Blood was collected via cardiac puncture and then centrifuged at 3000 rpm for 20 min to obtain serum samples. Additionally, tissues and fecal samples were collected and carefully preserved at −70 °C for future analyses.

2.8. Micro-Computed Tomography (CT) of the Femur

The femur was then fixed with 4% paraformaldehyde, and the BMD was determined in vivo through X-Ray Radiography Micro-computed tomography (Micro-CT; Skyscan 1273, Billerica, MA, USA) in Korea Basic Science Institute, Aging Science Center (Kwangju, Republic of Korea).

2.9. Biochemical Assay

The homeostasis model assessment for insulin resistance (HOMA-IR) was determined by applying the following formula: fasting insulin (μIU/mL) × fasting glucose (mM)/22.5. The Glucose Analyzer II (Beckman-Coulter, Palo Alto, CA, USA) and Ultrasensitive insulin ELISA kits (R&D Diagnostics, Minneapolis, MN, USA) were employed to measure the serum glucose and insulin concentrations, respectively. ELISA kits from Enzo Life Sciences (Farmingdale, NY, USA) were employed to assess serum 17β-estradiol levels. Additionally, BMD-related biomarkers in circulation were measured using ELISA kits. These biomarkers included rat osteoprotegerin (OPG; Abcam, Cambridge, UK), receptor activator of nuclear factors-κB ligand (RANKL; Abcam), osteocalcin (Abcam), parathyroid hormone (PTH; Abcam), and bone-specific alkaline phosphatase (BALP; MyBioSource, San Diego, CA, USA).

Serum was collected from portal vein blood and combined with acidic ethanol (0.01 N HCl; Duksan, Republic of Korea). SCFA concentrations in the resulting supernatants were measured using a gas chromatograph (Clarus 680 GAS, PerkinElmer, Waltham, MA, USA) equipped with an Elite-FFAP capillary column (30 m × 0.25 mm × 0.25 μm) [27]. External standards of 1 mM acetate, propionate, and butyrate (Sigma Co., St. Louis, MO, USA) were used for calibration.

2.10. Gut Microbiota in the TMD and Feces Evaluated by NGS

Metagenome sequencing using NGS techniques was conducted to analyze the microbial communities in TMD and fecal samples obtained from the cecum [10]. Bacterial DNA from the feces was extracted using the Power Water DNA Isolation Kit (Qiagen, Valencia, CA, USA). PCR products were amplified with 16S amplicon primers in the FastStart High-Fidelity PCR System (Roche, Basel, Switzerland), following the GS FLX plus library prep guide [10]. The bacterial DNA in the feces was then sequenced using Illumina MiSeq and a Genome Sequencer FLX plus 454 Life Sciences (Illumina; San Diego, CA, USA) following the manufacturer's instructions in Macrogen (Seoul, Republic of Korea) [28].

The 16S amplicon sequences were processed using Mothur v.1.36, and the Miseq standard operation procedure was applied to identify the taxonomy of fecal bacteria. Silva reference alignment v.12350 was utilized to align the sequences, and relative bacterial counts were determined based on taxonomic assignment for each sample [28]. The sequences classified as mitochondria, Eukaryota, or unknown were removed. Operational taxonomic units (OTUs) below 10,000 reads were deleted. The principal coordinate analysis (PCoA) outcomes of gut microbiota were visualized using the R package, as described previously [10].

Metagenome function analysis was conducted using the PICRUSt2 software to investigate the differences among gut microbiota in the various groups. The metabolic functions of the fecal bacteria were predicted from the FASTA files and counted using tables PICRUSt2, and these predictions were based on the Kyoto Encyclopedia of Genes and Genomes (KEGG) Orthologues (KO) mapped through the KEGG mapper (https://www.genome.jp/kegg/tool/mapper/search.html, accessed on 23 March 2023), as previously described [10].

2.11. Statistical Analysis

Statistical analysis was performed using SAS software version 7 (SAS Institute, Cary, NC, USA). The optimal sample size of 10 per group was determined using the G power program with a power of 0.90 and an effect size of 0.5. After confirming the normal distribution through Proc univariate, the data were expressed as mean ± standard deviation (SD). One-way ANOVA was employed to analyze the measurements, and Tukey's test was used to assess differences among the groups. Statistical significance was considered at $p < 0.05$.

3. Results

3.1. The Characteristics of Dried TMD with Different Amounts of B. subtilis and Biogenic Amines

Water content in the TMD samples was around 50% (50–59%), and the dried TMD contained approximately 7–9% salts (Table 1). The amounts of histamine and tyramine, biogenic amines, were higher in two TMDs (HSHA and LSHA) than in the others (Table 1). The biogenic amine contents of HSHA and LSHA were higher than HSLA and LSLA. The bacterial composition of the TMD products, as determined by NGS, exhibited that two HSHA and HSLA contained about 90% *B. subtilis* but other TMDs (LSHA and LSLA) were composed of less than 60%. LSHA and LSLA contained about 20 and 35% of *B. licheniformis. Leuconostoc mesenteroides*, *Leuconostoc citreum*, *Staphylococcus aureus*, and *Acinetobacter baumannii*, potentially harmful bacteria, were less than 1% in all TMDs (Table 1).

CSB contained high in isoflavonoid glycans and low in isoflavonoid glycans. After fermentation, isoflavonoid glycans were not detected in LSHA, HSHA, and HSLA, but LSLA contained a small number of isoflavonoid glycans. However, TMD included much higher amounts of isoflavonoid aglycans than CSB. Among TMDs, LSLA contained less isoflavonoid aglycon than LSHA, HSHA, and HSLA (Table 1). Total isoflavonoids were much lower in TMDs than CSB, and LSLA contained them the lowest.

Table 1. Characteristics of dried traditionally-made doenjang (TMD).

	LSHA	HSHA	HSLA	LSLA	CSB
Sodium (%)	8.8 ± 0.1 [a]	7.4 ± 0.2 [c]	9.0 ± 0.1 [a]	8.5 ± 0.2 [b]	-
Water (%)	56.7 ± 0.37 [b]	59.2 ± 0.32 [a]	51.9 ± 0.33 [c]	49.9 ± 0.36 [d]	
Histamine (ug/g)	796 ± 1.4 [b]	954 ± 0.6 [a]	22.7 ± 0.2 [d]	59.9 ± 0.4 [c]	-
Tyramine (ug/g)	2629 ± 1.4 [a]	1653 ± 1.4 [b]	36.1 ± 1.4 [d]	279 ± 1.7 [c]	-
Bacillus subtilis (%)	15.0 ± 0.6	86.5 ± 2.3	92.0 ± 1.6	59.4 ± 1.5	-
Bacillus licheniformis (%)	35.1 ± 0.72	0.60 ± 0.0	0 ± 0	19.7 ± 0.12	-
Weissella confuse (%)	0.02 ± 0	4.3 ± 0.2	0.35 ± 0.01	0.01 ± 0	-
Pediococcus acidilactici (%)	0.16 ± 0	2.06 ± 0.07	0.05 ± 0	0.02 ± 0	-
Bacillus coagulans (%)	0.76 ± 0.02	0 ± 0	0 ± 0	0.03 ± 0	-
Leuconostoc mesenteroides (%)	0 ± 0	0.9 ± 0.04	0.01 ± 0	0.15 ± 0	-
Leuconostoc citreum (%)	0 ± 0	0.39 ± 0.01	0 ± 0	0.01 ± 0	-
Staphylococcus aureus (%)	0.09 ± 0	0 ± 0	0 ± 0	0.01 ± 0	-
Acinetobacter baumannii (%)	0.01 ± 0	0.04 ± 0	0 ± 0	0.01 ± 0	-
Daidzein (ug/g)	25.9 ± 0.05 [b]	26.4 ± 0.62 [b]	29.9 ± 0.48 [a]	13.4 ± 0.31 [c]	8.4 ± 0.11 [d]
Genistein (ug/g)	43.8 ± 0.47 [b]	42.6 ± 0.54 [b]	47.2 ± 0.34 [a]	26.3 ± 0.38 [c]	9.2 ± 0.09 [b]
Glycitein (ug/g)	6.7 ± 0.11 [a]	6.3 ± 0.08 [a]	6.8 ± 0.9 [a]	1.8 ± 0.22 [b]	3.5 ± 0.17 [ab]
Daidzin (ug/g)	-	-	-	1.2 ± 0.09	35.4 ± 0.39
Genistin (ug/g)	-	-	-	2.5 ± 0.08	50.5 ± 0.58
Glycitin (ug/g)	-	-	-	-	8.7 ± 0.24
Total isoflavonoid aglycans (ug/g)	76.4 ± 1.87 [b]	75.3 ± 1.11 [b]	83.9 ± 0.73 [b]	45.2 ± 0.56 [c]	116 ± 0.82 [a]

Values represented means ± standard deviation (n = 5). HSLA, TMD with high contents of *Bacillus subtilis (B. subtilis)* and high biogenic amines. HSHA, TMD with high contents of *B. subtilis* and low biogenic amines. LSHA, TMD with low contents of *B. subtilis* and high biogenic amines. LSLA, TMD with low contents of *B. subtilis* and low biogenic amines. [a,b,c,d] Different letters on the bars indicate a significant difference among the groups by Tukey's test at $p < 0.05$.

3.2. Energy and Glucose Metabolism

Throughout the 12-week intervention, the Control group displayed a higher body weight increase than the Normal-control group. In contrast, all TMD groups exhibited lower weight gain than the Control group, but the reduction was not as significant as observed in the Normal-control group (Table 2). CSB did not result in decreased weight gain compared to the Control group. However, it is noteworthy that all groups showed no significant difference in food intake (Table 2).

Table 2. Serum glucose, insulin, and 17β-estradiol concentrations after 12-week intervention.

	Control	LSHA	HSHA	HSLA	LSLA	CSB	Normal-Con
Weight gain (g/12 week)	175 ± 12.1 [a]	150 ± 12.7 [b]	157 ± 10.1 [b]	152 ± 12.1 [b]	159 ± 14.2 [b]	165 ± 16.3 [ab]	117 ± 13.4 [c]
Visceral fat (g)	11.1 ± 0.95 [a]	10.8 ± 1.02 [a]	7.73 ± 0.84 [b]	7.35 ± 0.77 [b]	5.92 ± 0.63 [c]	7.15 ± 0.82 [b]	5.75 ± 0.73 [c]
Food intake (g/day)	13.6 ± 1.8	12.2 ± 1.3	12.6 ± 1.2	12.5 ± 1.4	13.0 ± 1.3	12.9 ± 1.4	11.6 ± 1.7
Uterine weight (g)	0.16 ± 0.02 [b]	0.15 ± 0.02 [b]	0.17 ± 0.03 [b]	0.15 ± 0.02 [b]	0.16 ± 0.02 [b]	0.15 ± 0.02 [b]	0.65 ± 0.03 [a]
Serum 17β-estradiol(pg/mL)	1.56 ± 0.10 [b]	1.52 ± 0.14 [b]	1.61 ± 0.25 [b]	1.49 ± 0.22 [b]	1.53 ± 0.28 [b]	1.42 ± 0.29 [b]	7.47 ± 0.88 [a]
Fasting serum glucose (mg/dL)	113 ± 4.47 [a]	104 ± 5.58 [ab]	103 ± 4.76 [b]	102 ± 5.34 [b]	109 ± 6.03 [ab]	109 ± 7.46 [ab]	95.3 ± 5.43 [c]
2h-postprandial serum glucose (mg/dL)	143 ± 7.23 [a]	126 ± 6.75 [b]	126 ± 6.55 [b]	126 ± 7.56 [b]	141 ± 7.32 [a]	135 ± 7.44 [ab]	125 ± 6.13 [b]
Fasting plasma insulin (ng/mL)	1.56 ± 0.19 [a]	1.11 ± 0.19 [b]	1.04 ± 0.13 [b]	0.83 ± 0.19 [c]	1.49 ± 0.21 [a]	0.93 ± 0.19 [bc]	0.86 ± 0.10 [c]
HOMA-IR	6.53 ± 0.53 [a]	4.93 ± 0.53 [b]	3.97 ± 0.34 [c]	3.11 ± 0.44 [c]	6.02 ± 0.59 [a]	3.75 ± 0.51 [c]	3.04 ± 0.26 [c]

Values represent mean ± standard deviation (n = 10). HOMA-IR, homeostasis model assessment estimate for assessing insulin resistance. HSLA, TMD with high contents of *Bacillus subtilis (B. subtilis)* and high biogenic amines. HSHA, TMD with high contents of *B. subtilis* and low biogenic amines. LSHA, TMD with low contents of *B. subtilis* and high biogenic amines. LSLA, TMD with low contents of *B. subtilis* and low biogenic amines. [a,b,c] Different letters in each variable indicate significant differences in one-way ANOVA at $p < 0.05$.

The Control group exhibited significantly lower serum 17β-estradiol concentrations and uterine weight than the Normal-control group. The administration of TMD or CSB did not affect these concentrations (Table 2). Uterine weight was notably higher in the Normal-control group than in the Control group, but TMD did not significantly impact it (Table 2). The fasting and 2-h post-prandial serum glucose concentrations were higher in the Control group than in the Normal-control group. HSLA prevented this increase. However, the 2-h post-prandial serum glucose concentrations in the HSHA and HSLA groups were similar to those of the Normal-control group (Table 2). Fasting serum insulin concentrations were higher in the Control group compared to the Normal-control group. Interestingly, HSHA, HSLA, and CSB intake led to decreased insulin concentrations. The serum insulin concentrations in the HSLA and CSB groups were similar to those of the Normal-control group (Table 2). HOMA-IR, an insulin resistance index, was significantly higher in the Control group than in the Normal-control group. However, HSHA, HSLA, and CSB intake resulted in similar levels of HOMA-IR compared to the Normal-control group (Table 2).

3.3. OGTT and IPITT

Following the administration of 2 g of glucose per kilogram of body weight in the OGTT, the serum glucose concentrations of all rats steadily increased up to 30–40 min and then gradually decreased (Figure 1A). At 20 min, the peak serum glucose concentrations did not show significant differences between the groups, but they markedly decreased between 30 and 50 min. Subsequently, the concentrations gradually decreased from 50 min onwards. The serum glucose concentrations in the Control group were higher than those in the Normal-control group, whereas they were lower in the LSHA and CSB groups. The AUC of the serum glucose concentrations from 0–40 min was higher in the Control group than the Normal-control group, while it was lower in all the TMD and CSB groups than in the Normal-control (Figure 1B). However, the AUC from 40–90 min was much higher in the Control group than the Normal-control group, while the AUC in the LSHA and CSB groups was similar to that of the Normal-control (Figure 1B).

During the OGTT, serum insulin concentrations in the Control group increased to 30 min but peaked at 20 min for the other groups. The Normal-control group showed the lowest serum insulin concentrations, with the HSHA and HSLA groups displaying levels similar to the Normal-control (Figure 1C). The AUC of the serum insulin concentrations during 0–20 min was higher in the Control group compared to the Normal-control group, and it was lower in the HSHA and HSLA groups than the Control group but higher than the Normal-control (Figure 1D). The AUC of the serum insulin concentrations from 20 to 90 min was much higher in the Control group compared to the Normal-control group. While it was lower in the HSHA group than the Control group, it was not as much as in the Normal-control group (Figure 1D).

Following intraperitoneal insulin injection after 6-h food deprivation in the IPITT, serum glucose concentrations decreased in all rats. However, the decrease was less significant in the Control group than in the other groups (Figure 2A). At 60–90 min, the concentrations in the HSHA, HSLA, and CSB groups were lower than in the Normal-control. The AUC of the serum glucose concentrations was much higher in the Control than in the Normal-control group (Figure 2A). Meanwhile, the AUC for the HSHA, HSLA, LSLA, and CSB groups decreased during the 0–30 min period. The AUC between 30 and 90 min also showed a similar decrease but to a greater extent (Figure 2B). This indicates that HSHA, HSLA, and CSB can improve glucose homeostasis and reduce insulin resistance.

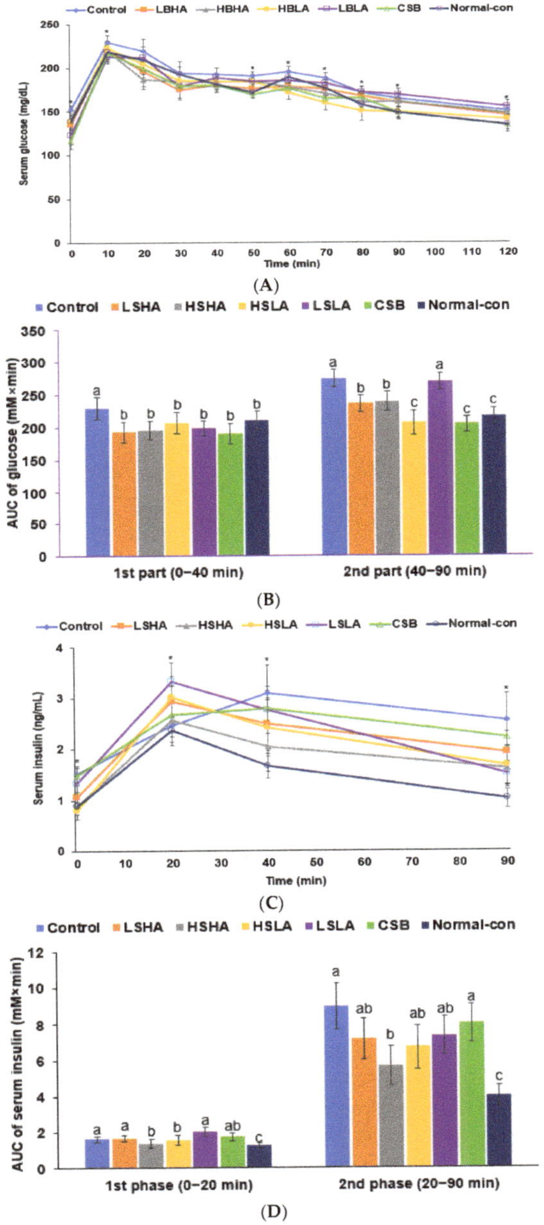

Figure 1. Serum glucose and insulin concentrations after the oral intake of 2 g glucose per kg body weight during oral glucose tolerance test (OGTT). (**A**) Changes in serum glucose concentrations. (**B**) The area under the curve (AUC) of serum glucose concentrations during the OGTT. (**C**) Changes in serum insulin concentrations. (**D**) The AUC of serum insulin concentration during the OGTT. Dots or bars and error bars represent the means ± standard deviations (n = 10). * significantly different among six groups at $p < 0.05$. [a,b,c] Different letters on the bars indicate a significant difference among the groups by Tukey's test at $p < 0.05$.

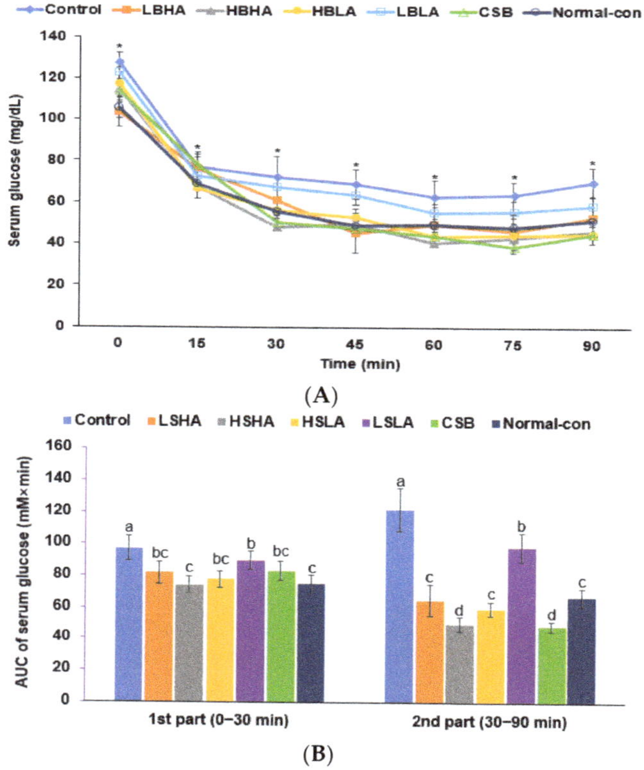

Figure 2. Serum glucose concentrations after the intraperitoneal injection of 1 U insulin per kg body weight during intraperitoneal insulin tolerance test (IPITT). (**A**) Changes in the serum glucose concentrations. (**B**) The area under the curve (AUC) of serum glucose concentrations during the IPITT. Dots or bars and error bars represent the means ± standard deviations (n = 10). * significantly different among six groups at $p < 0.05$. [a,b,c,d] Different letters on the bars indicated a significant difference among the groups by Tukey's test at $p < 0.05$.

3.4. Memory Functions and Depression

In the passive avoidance test, the latency in entering the dark room was significantly shorter in the Control group compared to the Normal-control group in the second trial, and it was only extended in the HSHA group. In the third trial, the latency was longer in all the TMD and CSB groups except LSLA, and it was similar to that of the Normal-control group (Figure 3A). The novel object recognition rate was lower in the Control group than in the Normal-control group during the novel object recognition test. However, administration of HSHA and HSLA to OVX rats increased the recognition rate, making it comparable to that of the Normal-control group (Figure 3B). The swimming rate assessed depression during the forced swimming test (Figure 3B). The forced swimming rate was lower in the Control group than the Normal-control group but increased in the HSHA and HSLA groups, reaching levels similar to that of the Normal-control group. The result suggests that OVX rats induced depression, and HSHA and HSLA treatments effectively ameliorated it, similar to the Normal-control group.

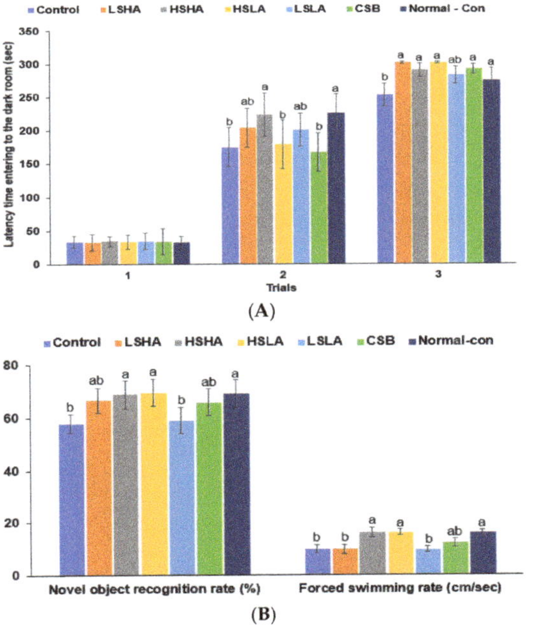

Figure 3. Cognitive function. (**A**) The latency in entering the dark room in the passive avoidance test. (**B**) Novel object recognition rate and forced swimming test by rats. Bars and error bars represent the means ± standard deviations (n = 10). [a,b] Different letters on the bars indicated a significant difference among the groups by Tukey's test at $p < 0.05$.

3.5. BMD by DEXA and Micro-CT

The DEXA analysis showed that the differences in the BMD of the lumbar spine and left and right legs between the pre- and post-interventions were lower in the Control group than in the Normal-control group based on measurements taken before and after the 12-week intervention (Figure 4A). In the HSHA and HSLA groups, there was no difference in the BMD of the lumbar spine after the interventions, similar to the Normal-control. The differences in BMD of the legs were higher than that of the lumbar spine, and the decrease in the HSHA group was similar to that of the Normal-control (Figure 4A). The difference in the BMD of the legs was lower in HSLA than in the Normal-control.

At the end of the 12-week intervention, the Micro-CT analysis revealed that the BMD in the Control group was significantly lower than in the Normal-control group (Figure 4B). However, all the TMD and CSB groups showed significantly higher BMD than the Control group. The HSLA group exhibited a remarkable prevention of BMD loss, although its BMD levels did not fully reach those observed in the Normal-control (Table 3, Figure 4B). The ratio of segmented bone volume to the total volume of the bone region (BV/TV) was higher in the Normal-control compared to the Control, and the TMD and CSB treatments effectively prevented the decrease in BV/TV. The HSLA group showed the highest BV/TV, though it still remained lower than the Normal-control (Table 3; Figure 4B). The mean trabecular thickness (Tb_Th) exhibited a similar trend to BV/TV (Table 3; Figure 4B). Moreover, the average trabecular number (Tb.N) was lower in the Control group than in the Normal-control group, and the HSLA group showed the highest Tb.N among the TMD and CSB treatment groups. On the other hand, the mean trabecular distance (Tb.Sp) displayed an opposite trend to that of Tb.N (Table 3; Figure 4B). Overall, both HSLA and HSHA treatments were effective in partially preventing BMD loss better than CSB in estrogen deficiency-induced conditions. However, they were unable to maintain BMD at the level of the Normal-control group fully.

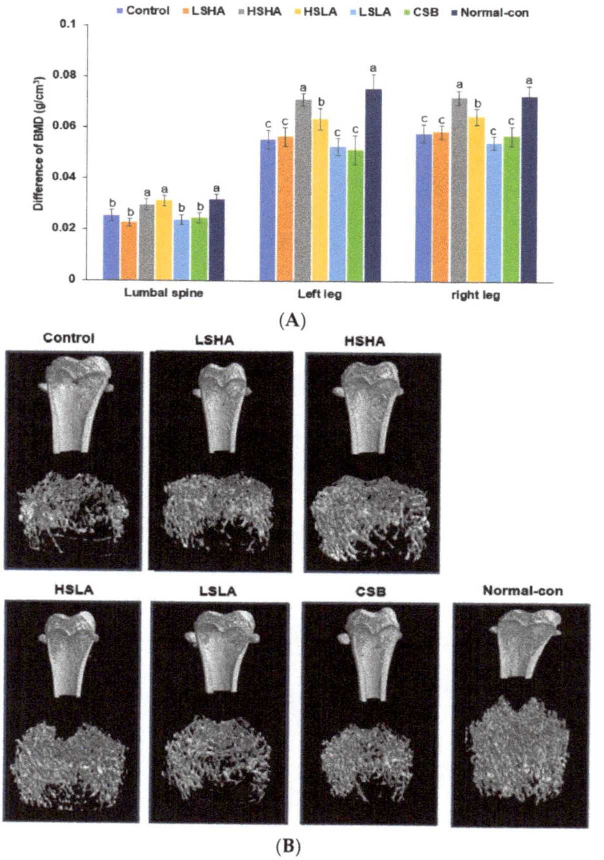

Figure 4. Bone mineral density measured by dual X-ray absorptiometry (DEXA) and Micro-computed tomography (CT). (**A**) Differences in the BMD before and after the 12-week intervention of TMD or CSB. (**B**) Image of micro-CT analysis in the distal femur after the 12-week treatment. Bars and error bars represent the means ± standard deviations (n = 10). [a,b,c] Different letters on the bars indicate a significant difference among the groups by Tukey's test at $p < 0.05$.

The serum PTH concentrations were notably higher in the Control group than in the Normal-control group (Table 4). However, the administration of CSB and TMD, except for LSHA and LSLA, resulted in decreased serum PTH concentrations compared to the Control group. Remarkably, HSHA administration led to a lower concentration of PTH, even comparable to the levels observed in the Normal-control group. Regarding the serum RANKL concentrations, they increased in the Control group when compared to the Normal-control group. However, the administration of TMD and CSB, except for LSHA and LSLA, decreased RANKL concentrations, bringing them to similar levels as in the Normal-control group (Table 4). As for the serum OPG concentrations, they were reduced in the Control group when compared to the Normal-control group. Nevertheless, the administration of HSHA, HSLA, and CSB effectively decreased the OPG levels, bringing them to levels similar to those observed in the Normal-control group (Table 4). Moreover, the serum osteocalcin concentrations were higher in the Control group than in the Normal-control group, but the administration of HSHA, HSLA, and CSB successfully prevented their increase (Table 4). As an index of bone turnover rate, serum BALP concentrations were higher in the Control group than in the Normal-control group. The administration of TMD and CSB prevented the increase in BALP concentrations. However, only the CSB group exhibited BALP concentrations similar to those in the Normal-control group (Table 4).

Table 3. Bone mineral density (BMD) of the femur by micro-CT after a 12-week intervention.

	Control	LSHA	HSHA	HSLA	LSLA	CSB	Normal-Con
BMD (g/cm^3)	0.065 ± 0.016 ^{d}	0.106 ± 0.013 ^{c}	0.122 ± 0.016 ^{c}	0.18 ± 0.047 ^{b}	0.081 ± 0.019 ^{d}	0.114 ± 0.001 ^{c}	0.23 ± 0.016 ^{a}
BV/TV (%)	11.4 ± 2.1 ^{e}	15.9 ± 1 ^{d}	17.9 ± 2.0 ^{c}	23.8 ± 3.0 ^{b}	13.5 ± 2.1 ^{e}	17.1 ± 1.3 ^{c}	30.6 ± 1.1 ^{a}
Tb.Th (mm)	0.096 ± 0.003 ^{c}	0.1 ± 0.005 ^{b}	0.106 ± 0.002 ^{a}	0.099 ± 0.001 ^{b}	0.098 ± 0.004 ^{b}	0.098 ± 0.003 ^{b}	0.111 ± 0.003 ^{a}
Tb.N (1/mm)	1.18 ± 0.08 ^{c}	1.6 ± 0.18 ^{b}	1.63 ± 0.19 ^{b}	2.39 ± 0.36 ^{a}	1.38 ± 0.24 ^{c}	1.74 ± 0.01 ^{b}	2.71 ± 0.08 ^{a}
Tb.Sp (mm)	1.12 ± 0.08 ^{a}	0.77 ± 0.19 ^{b}	0.81 ± 0.15 ^{b}	0.49 ± 0.21 ^{c}	1.02 ± 0.23 ^{a}	0.72 ± 0.05 ^{b}	0.4 ± 0.08 ^{c}

Values represent mean ± standard deviation (n = 10). HSLA, TMD with high contents of *Bacillus subtilis (B. subtilis)* and high biogenic amines. HSHA, TMD with high contents of *B. subtilis* and low biogenic amines. LSHA, TMD with low contents of *B. subtilis* and high biogenic amines. LSLA, TMD with low contents of *B. subtilis* and low biogenic amines. BMD, bone mineral density; BV/TV (%), bone volume/tissue volume; BS/BV, bone surface/bone volume; Tb.Th, trabecular thickness; Tb.N, trabecular number; Tb.Sp, trabecular separation. ^{a,b,c,d,e} Different letters in each variable indicate significant differences in one-way ANOVA at $p < 0.05$.

Table 4. Bone-metabolism-related parameters after the 12-week intervention.

	Control	LSHA	HSHA	HSLA	LSLA	CSB	Normal-Con
PTH (ng/mL)	43.9 ± 5.49 ^{a}	38.8 ± 4.62 ^{b}	28 ± 5.15 ^{c}	32 ± 6.7 ^{bc}	39.4 ± 5.03 ^{ab}	31.4 ± 6.32 ^{bc}	37.5 ± 3.89 ^{b}
RANKL (pg/mL)	32.3 ± 3.43 ^{a}	29.4 ± 3.15 ^{b}	28.1 ± 3.52 ^{b}	24.1 ± 3.12 ^{c}	30.9 ± 3.73 ^{ab}	27.9 ± 4.12 ^{b}	22.4 ± 3.52 ^{c}
OPG (ng/mL)	14.4 ± 0.95 ^{c}	14.7 ± 0.97 ^{c}	17.5 ± 0.83 ^{b}	16.9 ± 0.95 ^{b}	18.9 ± 1.17 ^{a}	17.2 ± 0.94 ^{b}	17.3 ± 0.91 ^{b}
Osteocalcin (ng/mL)	1.59 ± 0.17 ^{a}	1.40 ± 0.21 ^{ab}	1.13 ± 0.18 ^{b}	1.06 ± 0.17 ^{b}	1.38 ± 0.15 ^{ab}	1.05 ± 0.14 ^{b}	0.69 ± 0.08 ^{c}
BALP (U/L)	37.2 ± 4.58 ^{a}	28.2 ± 4.72 ^{b}	26.9 ± 3.68 ^{b}	25.4 ± 5.09 ^{b}	28 ± 5.01 ^{b}	19.9 ± 4.18 ^{c}	15.9 ± 3.2 ^{c}

Values represent mean ± standard deviation (n = 10). HSLA, TMD with high contents of *Bacillus subtilis (B.) subtilis* and high biogenic amines. HSHA, TMD with high contents of *B. subtilis* and low biogenic amines. LSHA, TMD with low contents of *B. subtilis* and high biogenic amines. LSLA, TMD with low contents of *B. subtilis* and low biogenic amines. PTH, parathyroid hormone; OPG, Osteoprotegerin; RANKL, receptor activity of nuclear factor kappa B ligand; BALP, bone-specific alkaline phosphatase. ^{a,b,c} Different letters in each variable indicate significant differences in one-way ANOVA at $p < 0.05$.

3.6. SCFA in the Blood from the Portal Vein and Fecal Bacteria

The levels of acetate and propionate in the portal vein did not show significant differences among all the groups (Figure 5A). However, butyrate concentrations were notably lower in the Control group compared to the Normal-control group. The HSLA and CSB groups exhibited the highest butyrate levels among the intervention groups, whereas the LSHA and LSLA groups had relatively lower levels (Figure 5A).

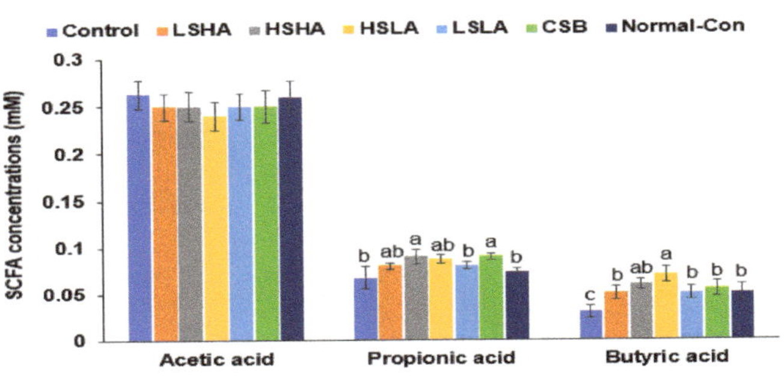

(A)

Figure 5. *Cont.*

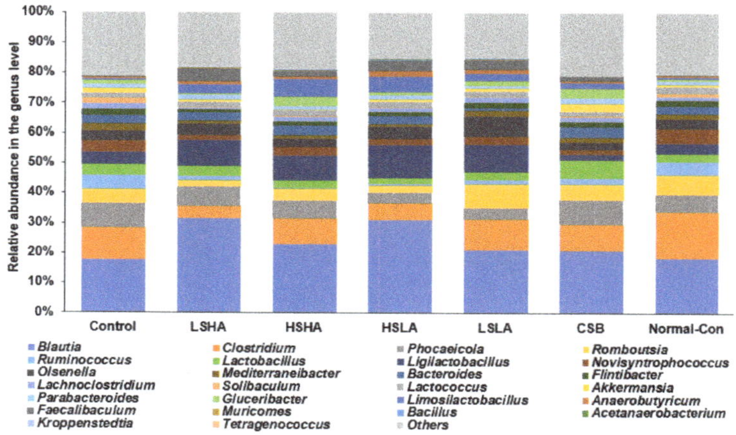

(**B**)

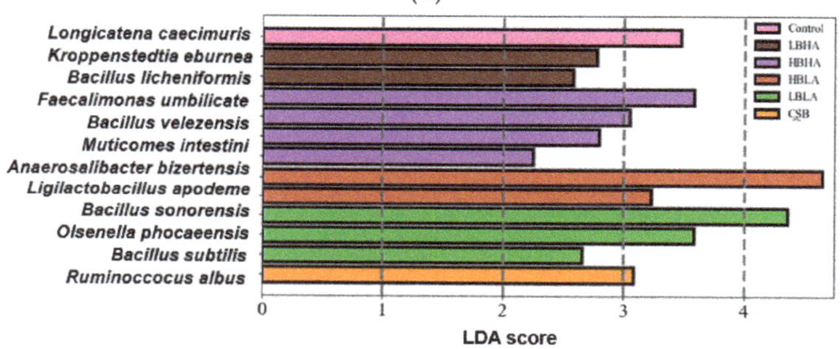

(**C**)

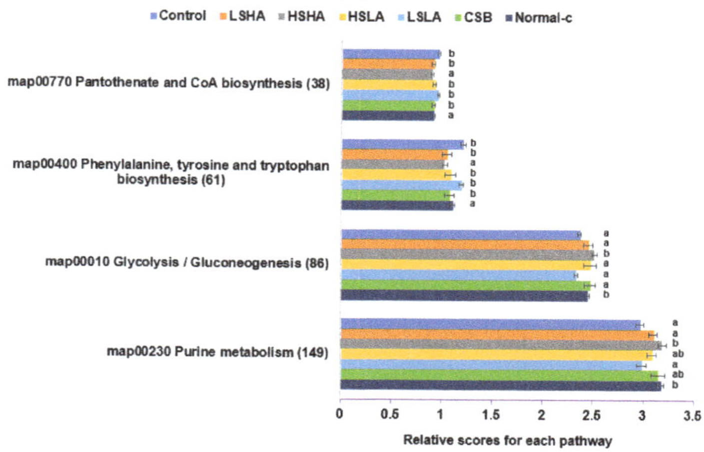

(**D**)

Figure 5. *Cont.*

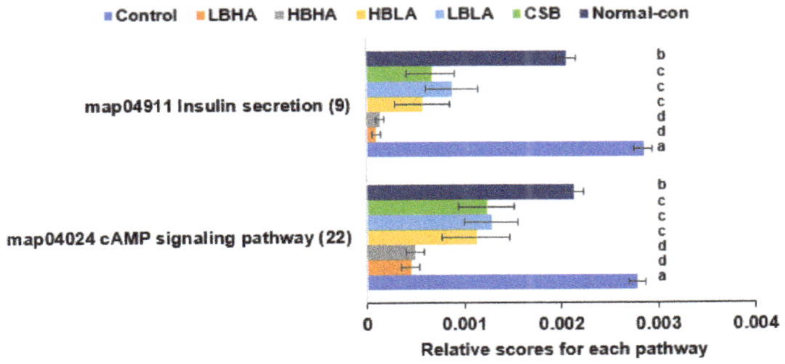

(E)

Figure 5. Fecal bacterial analysis after the 12-week intervention. (**A**) The short-chain fatty acid concentration in the portal vein. (**B**) Relative abundance of fecal bacteria from the cecum at the genus level. (**C**) Linear discriminant analysis (LDA) score. (**D**) Metagenome analysis in glucose, amino acid, and purine metabolism. (**E**) Metagenome analysis in insulin secretion and cyclic adenosine 3′,5′-monophosphate (cAMP) signaling pathways. Bars and error bars represent the means ± standard deviations ($n = 10$). [a,b,c,d] Different letters on the bars indicated a significant difference among the groups by Tukey's test at $p < 0.05$.

The Control group had somewhat different fecal bacteria than the Normal-control and the TMD intake altered their composition (Figure 5B). The levels of the genus *Blautia* were higher in the LSHA and HSLA groups than the other groups, and levels of the genus *Bacillus* were higher in the TMD groups than in the Control and Normal-control groups (Figure 5B). The TMD intake increased the abundance of the genus *Lactobacillus* and *Ligilactobacillus* compared to the Control and Normal-control groups, and the intake of LSLA increased the abundance of the genus *Olsenella* compared to the other groups (Figure 5B). Although the α-diversity did not vary among the groups, TMD was separated from the Control in the β-diversity ($p < 0.001$). In linear discriminant analysis (LDA) analysis, *Bacillus* species were selected as the primary bacteria in the TMD but not the CSB group. *Ligilactobacillus apodemi* had a high LDA level in the HSLA group, and *Olsenella phocaeensis* and *Clostridium hylemonae* were high in the LSHA group (Figure 5C). Regarding metagenome function analysis, glycolysis/gluconeogenesis and purine metabolism were lower in the Control group compared to the Normal-control group. CSB and TMD, except LSLA, prevented the decrease in these functions. Pantothenate and coenzyme A (CoA) biosynthesis and phenylalanine/tyrosine/tryptophan biosynthesis showed an opposite trend to glycolysis/gluconeogenesis in fecal bacteria (Figure 5D). Insulin secretion and the cyclic adenosine 3′,5′-monophosphate (cAMP) signaling pathways were higher in the Control group compared to the Normal-control group. TMD and CSB prevented an increase in the signaling of these pathways (Figure 5E).

4. Discussion

TMD has been reported to ameliorate obesity, hypertension, neuroinflammation, neurodegeneration, and oxidative stress [22]. However, few studies have been reported to evaluate TMD effects in improving estrogen-deficiency-associated symptoms, glucose, energy, and bone metabolism [29]. We aimed to investigate how different TMD types modulate glucose, bone metabolism, and memory function in OVX rats, an estrogen-deficient animal model. The novel findings of the current study are as follows. The present study showed that TMD decreased serum glucose concentrations during OGTT and improved insulin resistance, which helped prevent glucose-induced bone loss and memory dysfunction. The HSHA and HSLA types of TMD were particularly effective in preventing the reduction in BMD and memory function. However, the study also found that the abundance of *Bacillus* in TMD may affect its efficacy for glucose metabolism, bone density, and memory function. Further research is needed to evaluate the optimal TMD type to achieve the abovementioned benefits.

Furthermore, fermented soybeans such as chungkookjang and TMD improve glucose metabolism [30]. Both contain isoflavone aglycones more than isoflavone glycones compared to unfermented soybeans, and the process to remove glucose from the isoflavone glycones is conducted with *Bacillus* species, particularly *B. subtilis*, *B. amyloliquefaciens*, and *B. licheniformis* in chungkookjang and TMD. Therefore, adults showing a better response to soybeans may have specific intestinal microbes (e.g., *Lactobacillus intestinalis* and *Lactobacillus johnsonii*) that convert daidzein into equol, a potent phytoestrogen, potentially to alleviate estrogen-deficient symptoms [17]. These intestinal microbes exist in those consuming soybean products, especially fermented soybeans. The present study showed that all TMD types decreased glucose intolerance, mainly by enhancing insulin sensitivity. Therefore, TMD can improve glucose tolerance and insulin resistance, and among the various TMD types, HSLA was most effective in regulating glucose homeostasis by enhancing insulin sensitivity in estrogen-deficient rats. Isoflavonoids in HSLA might have a higher chance of being converted into equol, which is potent for preventing and ameliorating BMD loss. It needs to have further study.

There is a close relationship between glucose and bone metabolism and cognitive function, and estrogen deficiency adversely affects all three parameters. Osteoporosis has been linked to cognitive decline and memory impairment, likely due to blood flow changes and brain oxygen delivery [31]. Consistent with previous studies [31], in the present study, an increase in serum PTH concentrations was observed in estrogen-deficient rats, which increased bone resorption, leading to decreased BMD as measured by the DEXA and micro-CT methods and elevated bone resorption indexes such as RANKL, osteocalcin, and BALP. HSLA and HSHA prevented the decrement of BMD compared to the Control group. CSB declined bone resorption compared to the Control group, but the HSLA and HSHA groups exhibited lower bone resorption than the CSB group. Although the relationship between soybean intake and BMD remains controversial in estrogen-deficient states, dietary isoflavonoids attenuate menopause-induced osteoporotic bone loss by decreasing bone resorption and stimulating bone formation [32]. Isoflavonoids decrease bone resorption through the OPG/RANKL/RANK pathways, similar to the HSLA and HSHA activity observed in the present study [31]. Previous research has shown that consuming fermented soybean products like natto and doenjang is associated with increased BMD in postmenopausal women [33–35]. In the current study, the intake of HSLA and HSHA, among various types of TMDs, effectively prevented BMD loss in OVX rats by reducing bone resorption by decreasing serum PTH concentration. These findings strongly suggest that the presence of high B. subtilis in TMD plays a critical role in mitigating BMD loss. This observation aligns with other studies that have highlighted the positive impact of B. subtilis C-3102 on BMD improvement in postmenopausal women through mechanisms such as inhibiting bone resorption, enhancing fecal Bifidobacterium, and reducing Fusobacterium [36]. Therefore, HSLA and HSHA act as synbiotics to improve BMD in an estrogen-deficient state.

The decrease in glucose metabolism with insulin resistance contributes to reduced synaptic plasticity [37], suggesting brain insulin resistance leads to cognitive decline. Estrogen deficiency is also involved in exacerbating insulin resistance in the brain, possibly disturbing cognitive function [38]. Postmenopausal women who consume soy isoflavones have better cognitive function compared to those who do not consume soybeans [38]. Both unfermented and fermented soybeans (chungkookjang and doenjang) improve memory function in an estrogen-deficient state [16,39–41]. However, long-term fermented soy products may have a greater effect on improving cognitive dysfunction, particularly memory, due to *Bacillus* species and changes in the soybean components, such as increased isoflavone aglycones, as observed in previous studies [40,41]. The present study has demonstrated that the memory function assessed using passive avoidance and novel object recognition tests was improved in HSLA and HSHA-treated estrogen-deficient rats compared to the Control group. The improvement in HSLA and HSHA-treated rats was greater than in CSB-treated rats. HSLA and HSHA contain an abundance of *Bacillus* species, primarily *B. subtilis, B. coagulans, B. amyloliquefaciens, and B. velezensis* that produce short-chain fatty acids, especially butyrate and propionate, similar to chungkookjang [39–41]. Therefore, better memory function in fermented soybeans is potentially related to the *Bacillus* species and their metabolites.

The composition of fecal bacteria can exhibit variations compared to that of the gut, as observed in previous studies [42]. Factors such as transit time through the gut, interactions with host cells and other bacteria, and environmental influences contribute to these differences in fecal bacterial composition. Furthermore, the abundance of bacterial groups in feces is known to be influenced by factors such as diet, age, and health status [42,43]. However, whether specific bacterial species are more likely to be excreted in the feces is still unclear. In our present study, we found that an increased intake of *B. subtilis* from TMD led to higher excretion of *Bacillus* species in the feces, while the levels of *Akkermencia* were relatively lower in fecal bacteria. TMD in *B. subtilis* might modulate other gut bacteria, including *Bacillus* species, leading to altering the fecal bacteria. HSLA and HSHA did not increase *Bacillus* species in the feces, suggesting that *B. subtilis* and other *Bacillus* species might increase in the gut. Interestingly, *B. subtilis* was one of the primary fecal bacteria in the LSLA group, indicating that *B. subtilis* in the LSLA might not reside in the gut. Additionally, TMD intake increased not only the abundance of *Bacillus* species but also *Lactobacillus* and *Ligilactobacillus* species in the feces compared to the Control and CSB groups. Our previous study demonstrated that a comparable TMD intake resulted in an increase in *Akkermencia* but a decrease in *Ligilactobacillus* in the cecal bacteria of the HSLA and LSLA groups compared to the Control group. These findings highlight the differences in bacterial content between cecal and fecal samples after the intervention and emphasize the need to interpret fecal bacterial data carefully. Considering the findings from our previous and present studies, TMD exhibits synbiotic properties and shows potential for beneficial effects on bone metabolism.

Fermented foods can sometimes contain toxins produced by specific bacteria. Among the common toxins are mycotoxins, including aflatoxin B1, bacterial toxins such as botulinum and amylosin, as well as biogenic amines [44,45]. Aflatoxins, which Aspergillus generates, are heat-stable and remain unaffected by ordinary cooking methods. Reducing aflatoxin levels in fermented foods is important, and certain strains of *B. subtilis*, *B. amyloliquefaciens*, and *B. lincheniformis* have been found to degrade aflatoxins effectively [45–47]. Considering TMD, which contains a high amount of *B. subtilis*, the strains in the TMD can help degrade aflatoxins during fermentation, thus reducing their presence in the final product. However, it should be noted that certain *B. amyloliquefaciens* strains have been reported to produce amylosin, a heat-stable compound that can stimulate cytokines from human macrophages [46]. Biogenic amines, such as histamine and tyramine, present in some fermented foods, can potentially lead to adverse effects [18,19]. Nevertheless, some strains of *B. subtilis* have the ability to degrade these biogenic amines during the fermentation process, which contributes to improved food safety. The present study investigated the levels of biogenic amines in TMD enriched with *B. subtilis* and found that they did not induce harmful effects. It suggests that the presence of *B. subtilis* might contribute to reducing the production of certain toxins during fermentation. However, further research is required to isolate specific strains of *B. subtilis* that can effectively reduce toxins during the fermentation of TMD.

There were some limitations of the present study. (1) The study used ovariectomized (OVX) rats as a model for menopausal women. While animal models provide valuable insights, there may be differences in responses between animals and humans, limiting the direct applicability of the findings to human populations. (2) The intervention period was limited to 12 weeks. Long-term effects and safety considerations of prolonged TMD consumption were not explored, and additional studies with extended follow-up periods are necessary. (3) The study suggests potential benefits of *B. subtilis*-rich TMD for menopausal women. However, this study did not establish specific dosages and optimal formulations of TMD for human consumption, warranting further investigation. (4) The study relied on an animal model, and extrapolating the results to human populations requires caution.

In conclusion, menopause is characterized by typical menopausal symptoms and insulin resistance. Exacerbation of insulin resistance during menopause leads to the dysregulation of glucose metabolism, bone loss, and memory dysfunction. This study aimed to examine the potential of TMD intake to modulate glucose and bone homeostasis and memory function in OVX rats. The findings revealed that consuming different types of TMD, distinguished by the abundance of *B. subtilis* and biogenic amines, resulted in improved insulin resistance, reduced serum glucose concentrations, and mitigated memory decline compared to the Control group. Notably, HSHA and HSLA formulations exhibited the ability to preserve BMD and counteract markers of osteoclastic activity. However, it is essential to acknowledge that the BMD of the TMD-treated groups did not reach the levels observed in the Normal-control group. Consequently, TMD enriched with high *B. subtilis* content may be recommended for enhancing glucose metabolism and partially improving BMD and memory function. More long-term animal studies to examine the effects over extended periods would offer valuable insights into the potential sustainability and long-lasting benefits of incorporating TMD into the diet. Further randomized clinical trials with a large number of postmenopausal women are warranted to validate the potential benefits of TMD and to provide evidence-based recommendations.

Author Contributions: Conceptualization, S.P. and D.-Y.J.; methodology, H.-J.Y., S.-J.J. and M.-S.R.; resources, X.W. and H.-J.Y.; data collection and analysis, T.Z., Y.Y., S.-J.J., M.-S.R. and C.L.; writing—original draft preparation, S.P.; writing—review and editing, T.Z., D.-Y.J. and H.-J.Y.; supervision, D.-Y.J., S.P. and H.-J.Y. All authors have read and agreed to the published version of the manuscript.

Funding: This work was supported by "functional research of fermented soybean food (safety monitoring)" under the Ministry of Agriculture, Food and Rural Affairs and partly Korea Agro-Fisheries and Food trade corporation.

Institutional Review Board Statement: This study was conducted according to the National Institute of Health Guide for the Care and Use of Laboratory Animals with the approval of the Hoseo University Animal Care and Use Committee (HSIACUC-22-03).

Informed Consent Statement: Not applicable.

Data Availability Statement: The data used to support the findings of this study can be made available by the corresponding author upon request.

Conflicts of Interest: The authors declare no conflict of interest.

References

1. Kronenberg, F. Menopausal Hot Flashes: A Review of Physiology and Biosociocultural Perspective on Methods of Assessment 1,2. *J. Nutr.* **2010**, *140*, 1380S–1385S. [CrossRef]
2. Haines, C.J.; Xing, S.-M.; Park, K.-H.; Holinka, C.F.; Ausmanas, M.K. Prevalence of menopausal symptoms in different ethnic groups of Asian women and responsiveness to therapy with three doses of conjugated estrogens/medroxyprogesterone acetate: The Pan-Asia menopause (PAM) study. *Maturitas* **2005**, *52*, 264–276. [CrossRef]
3. Alemany, M. Estrogens and the regulation of glucose metabolism. *World J. Diab.* **2021**, *12*, 1622–1654. [CrossRef] [PubMed]
4. Gregorio, K.C.R.; Laurindo, C.P.; Machado, U.F. Estrogen and Glycemic Homeostasis: The Fundamental Role of Nuclear Estrogen Receptors ESR1/ESR2 in Glucose Transporter GLUT4 Regulation. *Cells* **2021**, *10*, 99. [CrossRef] [PubMed]
5. Yang, J.; Hong, N.; Shim, J.-S.; Rhee, Y.; Kim, H.C. Association of Insulin Resistance with Lower Bone Volume and Strength Index of the Proximal Femur in Nondiabetic Postmenopausal Women. *J. Bone Metab.* **2018**, *25*, 123–132. [CrossRef] [PubMed]
6. Wang, L.; Zhang, D.; Xu, J. Association between the Geriatric Nutritional Risk Index, bone mineral density and osteoporosis in type 2 diabetes patients. *J. Diabetes Investig.* **2020**, *11*, 956–963. [CrossRef] [PubMed]
7. Conde, D.M.; Verdade, R.C.; Valadares, A.L.R.; Mella, L.F.B.; Pedro, A.O.; Costa-Paiva, L. Menopause and cognitive impairment: A narrative review of current knowledge. *World J. Psychiatry* **2021**, *11*, 412–428. [CrossRef]

8. Willmann, C.; Brockmann, K.; Wagner, R.; Kullmann, S.; Preissl, H.; Schnauder, G.; Maetzler, W.; Gasser, T.; Berg, D.; Eschweiler, G.W.; et al. Insulin sensitivity predicts cognitive decline in individuals with prediabetes. *BMJ Open Diabetes Res. Care* **2020**, *8*, e001741. [CrossRef]
9. Mosconi, L.; Berti, V.; Dyke, J.; Schelbaum, E.; Jett, S.; Loughlin, L.; Jang, G.; Rahman, A.; Hristov, H.; Pahlajani, S.; et al. Menopause impacts human brain structure, connectivity, energy metabolism, and amyloid-beta deposition. *Sci. Rep.* **2021**, *11*, 10867. [CrossRef]
10. Park, S.; Kim, D.S.; Kang, E.S.; Kim, D.B.; Kang, S. Low-dose brain estrogen prevents menopausal syndrome while maintaining the diversity of the gut microbiomes in estrogen-deficient rats. *Am. J. Physiol. Endocrinol. Metab.* **2018**, *315*, E99–E109. [CrossRef]
11. Carpenter, J.; Gass, M.L.; Maki, P.M.; Newton, K.M.; Pinkerton, J.V.; Taylor, M.; Utian, W.H.; Schnatz, P.F.; Kaunitz, A.M.; Shapiro, M.; et al. Nonhormonal management of menopause-associated vasomotor symptoms: 2015 position statement of The North American Menopause Society. *Menopause* **2015**, *22*, 1155–1172.
12. Jayusman, P.A.; Nasruddin, N.S.; Baharin, B.; Ibrahim, N.I.; Ahmad Hairi, H.; Shuid, A.N. Overview on postmenopausal osteoporosis and periodontitis: The therapeutic potential of phytoestrogens against alveolar bone loss. *Front. Pharmacol.* **2023**, *14*, 1120457. [CrossRef] [PubMed]
13. Tang, S.; Du, Y.; Oh, C.; No, J. Effects of Soy Foods in Postmenopausal Women: A Focus on Osteosarcopenia and Obesity. *J. Obes. Metab. Syndr.* **2020**, *29*, 180–187. [CrossRef]
14. Park, S.; Kim, D.S.; Kang, S.; Moon, B.R. Fermented soybeans, Chungkookjang, prevent hippocampal cell death and β-cell apoptosis by decreasing pro-inflammatory cytokines in gerbils with transient artery occlusion. *Exp. Biol. Med.* **2016**, *241*, 296–307. [CrossRef]
15. Khosravi, A.; Razavi, S.H. Therapeutic effects of polyphenols in fermented soybean and black soybean products. *J. Funct. Foods* **2021**, *81*, 104467. [CrossRef]
16. Lee, D.-H.; Kim, M.J.; Park, S.-H.; Song, E.-J.; Nam, Y.-D.; Ahn, J.; Jang, Y.-J.; Ha, T.-Y.; Jung, C.H. Bioavailability of Isoflavone Metabolites After Korean Fermented Soybean Paste (Doenjang) Ingestion in Estrogen-Deficient Rats. *J. Food Sci.* **2018**, *83*, 2212–2221. [CrossRef]
17. Heng, Y.; Kim, M.J.; Yang, H.-J.; Kang, S.; Park, S. Lactobacillus intestinalis efficiently produces equol from daidzein and chungkookjang, short-term fermented soybeans. *Arch. Microbiol.* **2019**, *201*, 1009–1017. [CrossRef] [PubMed]
18. Jeong, S.J.; Ryu, M.S.; Yang, H.J.; Wu, X.H.; Jeong, D.Y.; Park, S.M. Bacterial Distribution, Biogenic Amine Contents, and Functionalities of Traditionally Made Doenjang, a Long-Term Fermented Soybean Food, from Different Areas of Korea. *Microorganisms* **2021**, *9*, 1348. [CrossRef] [PubMed]
19. Eom, J.S.; Seo, B.Y.; Choi, H.S. Biogenic Amine Degradation by Bacillus Species Isolated from Traditional Fermented Soybean Food and Detection of Decarboxylase-Related Genes. *J. Microbiol. Biotechnol.* **2015**, *25*, 1519–1527. [CrossRef] [PubMed]
20. Mah, J.H.; Park, Y.K.; Jin, Y.H.; Lee, J.H.; Hwang, H.J. Bacterial Production and Control of Biogenic Amines in Asian Fermented Soybean Foods. *Foods* **2019**, *8*, 85. [CrossRef]
21. Cha, Y.S.; Park, Y.; Lee, M.; Chae, S.W.; Park, K.; Kim, Y.; Lee, H.S. Doenjang, a Korean fermented soy food, exerts antiobesity and antioxidative activities in overweight subjects with the PPAR-γ2 C1431T polymorphism: 12-week, double-blind randomized clinical trial. *J. Med. Food* **2014**, *17*, 119–127. [CrossRef] [PubMed]
22. Linares, D.M.; Ross, P.; Stanton, C. Beneficial Microbes: The pharmacy in the gut. *Bioengineered* **2016**, *7*, 11–20. [CrossRef]
23. Jeong, S.; Shin, M.; Jeong, S.; Yang, H.; Jeong, D. Characteristic analysis and production of short-ripened Korean traditional soy sauce added with rice bran. *J. Korean Soc. Food Sci. Nutr.* **2014**, *43*, 550–556. [CrossRef]
24. Reeves, P.G. Components of the AIN-93 diets as improvements in the AIN-76A diet. *J. Nutr.* **1997**, *127*, 838s–841s. [CrossRef]
25. Zhang, T.; Kim, M.J.; Kim, M.J.; Wu, X.; Yang, H.J.; Yuan, H.; Huang, S.; Yoon, S.M.; Kim, K.N.; Park, S. Long-Term Effect of Porcine Brain Enzyme Hydrolysate Intake on Scopolamine-Induced Memory Impairment in Rats. *Int. J. Mol. Sci.* **2022**, *23*, 3361. [CrossRef] [PubMed]
26. Cui, Y.; Sun, K.; Xiao, Y.; Li, X.; Mo, S.; Yuan, Y.; Wang, P.; Yang, L.; Zhang, R.; Zhu, X. High-salt diet accelerates bone loss accompanied by activation of ion channels related to kidney and bone tissue in ovariectomized rats. *Ecotoxicol. Environ. Saf.* **2022**, *244*, 114024. [CrossRef]
27. Park, S.; Yuan, H.; Zhang, T.; Wu, X.; Huang, S.K.; Cho, S.M. Long-term silk peptide intake promotes skeletal muscle mass, reduces inflammation, and modulates gut microbiota in middle-aged female rats. *Biomed. Pharmacother.* **2021**, *137*, 111415. [CrossRef] [PubMed]
28. Wu, X.; Kim, M.J.; Yang, H.J.; Park, S. Chitosan alleviated menopausal symptoms and modulated the gut microbiota in estrogen-deficient rats. *Eur. J. Nutr.* **2021**, *60*, 1907–1919. [CrossRef]
29. Zhang, T.; Yue, Y.; Jeong, S.J.; Ryu, M.S.; Wu, X.; Yang, H.J.; Li, C.; Jeong, D.Y.; Park, S. Improvement of Estrogen Deficiency Symptoms by the Intake of Long-Term Fermented Soybeans Rich in Bacillus Species through Modulating Gut Microbiota in Estrogen-Deficient Rats. *Foods* **2023**, *12*, 1143. [CrossRef]
30. Jeong, S.Y.; Jeong, D.Y.; Kim, D.S.; Park, S. Chungkookjang with High Contents of Poly-γ-Glutamic Acid Improves Insulin Sensitizing Activity in Adipocytes and Neuronal Cells. *Nutrients* **2018**, *10*, 1588. [CrossRef]
31. Sitges-Serra, A.; Girvent, M.; Pereira, J.A.; Jimeno, J.; Nogués, X.; Cano, F.J.; Sancho, J.J. Bone mineral density in menopausal women with primary hyperparathyroidism before and after parathyroidectomy. *World J. Surg.* **2004**, *28*, 1148–1152. [CrossRef] [PubMed]

32. Zheng, X.; Lee, S.K.; Chun, O.K. Soy Isoflavones and Osteoporotic Bone Loss: A Review with an Emphasis on Modulation of Bone Remodeling. *J. Med. Food* **2016**, *19*, 1–14. [CrossRef]
33. Słupski, W.; Jawień, P.; Nowak, B. Botanicals in Postmenopausal Osteoporosis. *Nutrients* **2021**, *13*, 1609. [CrossRef] [PubMed]
34. Ikeda, Y.; Iki, M.; Morita, A.; Kajita, E.; Kagamimori, S.; Kagawa, Y.; Yoneshima, H. Intake of fermented soybeans, natto, is associated with reduced bone loss in postmenopausal women: Japanese Population-Based Osteoporosis (JPOS) Study. *J. Nutr.* **2006**, *136*, 1323–1328. [CrossRef] [PubMed]
35. Seol, J.Y.; Youn, Y.N.; Koo, M.; Kim, H.J.; Choi, S.Y. Influence of water-soluble extracts of long-term fermented Doenjang on bone metabolism bioactivity and breast cancer suppression. *Food Sci. Biotechnol.* **2016**, *25*, 517–524. [CrossRef] [PubMed]
36. Takimoto, T.; Hatanaka, M.; Hoshino, T.; Takara, T.; Tanaka, K.; Shimizu, A.; Morita, H.; Nakamura, T. Effect of Bacillus subtilis C-3102 on bone mineral density in healthy postmenopausal Japanese women: A randomized, placebo-controlled, double-blind clinical trial. *Biosci. Microbiota Food Health* **2018**, *37*, 87–96. [CrossRef] [PubMed]
37. Spinelli, M.; Fusco, S.; Grassi, C. Brain Insulin Resistance and Hippocampal Plasticity: Mechanisms and Biomarkers of Cognitive Decline. *Front. Neurosci.* **2019**, *13*, 788. [CrossRef]
38. Marchant, I.C.; Chabert, S.; Martínez-Pinto, J.; Sotomayor-Zárate, R.; Ramírez-Barrantes, R.; Acevedo, L.; Córdova, C.; Olivero, P. Estrogen, Cognitive Performance, and Functional Imaging Studies: What Are We Missing About Neuroprotection? *Front. Cell Neurosci.* **2022**, *16*, 866122. [CrossRef]
39. Jeong, D.Y.; Jeong, S.Y.; Zhang, T.; Wu, X.; Qiu, J.Y.; Park, S. Chungkookjang, a soy food, fermented with Bacillus amyloliquefaciens protects gerbils against ischemic stroke injury and post-stroke hyperglycemia. *Food Res. Int.* **2020**, *128*, 108769. [CrossRef]
40. Jeong, D.Y.; Ryu, M.S.; Yang, H.J.; Park, S. γ-PGA-Rich Chungkookjang, Short-Term Fermented Soybeans: Prevents Memory Impairment by Modulating Brain Insulin Sensitivity, Neuro-Inflammation, and the Gut-Microbiome-Brain Axis. *Foods* **2021**, *10*, 221. [CrossRef]
41. Zhang, T.; Ryu, M.S.; Wu, X.; Yang, H.J.; Jeong, S.J.; Seo, J.W.; Jeong, D.Y.; Park, S. Alleviation of Neuronal Cell Death and Memory Deficit with Chungkookjang Made with Bacillus amyloliquefaciens and Bacillus subtilis Potentially through Promoting Gut-Brain Axis in Artery-Occluded Gerbils. *Foods* **2021**, *10*, 2697. [CrossRef] [PubMed]
42. Zhou, X.; Zhang, S.; Liu, D.; Qian, H.; Zhang, D.; Liu, Q. The differences between fecal microbiota and intestinal fluid microbiota in colon polyps: An observational study. *Medicine* **2021**, *100*, e28028. [CrossRef] [PubMed]
43. Hasan, N.; Yang, H. Factors affecting the composition of the gut microbiota, and its modulation. *PeerJ* **2019**, *7*, e7502. [CrossRef] [PubMed]
44. Rasimus-Sahari, S.; Teplova, V.V.; Andersson, M.A.; Mikkola, R.; Kankkunen, P.; Matikainen, S.; Gahmberg, C.G.; Andersson, L.C.; Salkinoja-Salonen, M. The peptide toxin amylosin of Bacillus amyloliquefaciens from moisture-damaged buildings is immunotoxic, induces potassium efflux from mammalian cells, and has antimicrobial activity. *Appl. Environ. Microbiol.* **2015**, *81*, 2939–2949. [CrossRef]
45. Guo, Y.; Zhao, L.; Ma, Q.; Ji, C. Novel strategies for degradation of aflatoxins in food and feed: A review. *Food Res. Int.* **2021**, *140*, 109878. [CrossRef] [PubMed]
46. Afsharmanesh, H.; Perez-Garcia, A.; Zeriouh, H.; Ahmadzadeh, M.; Romero, D. Aflatoxin degradation by Bacillus subtilis UTB1 is based on production of an oxidoreductase involved in bacilysin biosynthesis. *Food Control* **2018**, *94*, 48–55. [CrossRef]
47. Chen, G.; Fang, Q.A.; Liao, Z.; Xu, C.; Liang, Z.; Liu, T.; Zhong, Q.; Wang, L.; Fang, X.; Wang, J. Detoxification of Aflatoxin B1 by a Potential Probiotic Bacillus amyloliquefaciens WF2020. *Front. Microbiol.* **2022**, *13*, 891091. [CrossRef]

foods

MDPI

Communication

Effects of Cheonggukjang (Fermented Soybean) on the Development of Colitis-Associated Colorectal Cancer in Mice

Hyeon-Ji Lim [1], In-Sun Park [1], Su-Ji Jeong [2], Gwang-Su Ha [2], Hee-Jong Yang [2], Do-Youn Jeong [2], Seon-Young Kim [1] and Chan-Hun Jung [1,*]

[1] Jeonju AgroBio-Materials Institute, Wonjangdong-gil 111-27, Jeonju 54810, Republic of Korea
[2] Microbial Institute for Fermentation Industry, Sunchang 56048, Republic of Korea
* Correspondence: biohun@gmail.com; Tel.: +82-63-711-1026

Abstract: Colorectal cancer (CRC) is the third most common type of cancer and is caused by multiple factors. Chronic inflammation, known to cause inflammatory bowel disease (IBD), is closely associated with CRC. Cheonggukjang (CJ), a traditional Korean fermented soybean, is a functional food with anti-inflammatory effects in the intestines, but its anti-cancer effects have not yet been explored. In this study, we investigated the cancer-protective effects of cheonggukjang in an azoxymethane/DSS (AOM/DSS)-induced colitis-associated colorectal cancer (CAC) mouse model. The CJ alleviated AOM/DSS-induced pathological symptoms such as colonic shortening, increased spleen weight, tumor formation, and histological changes. It also modulated pro-inflammatory and anti-inflammatory cytokine levels via the suppression of NF-κB and inflammatory mediator signaling pathways. Furthermore, the CJ improved intestinal integrity by regulating mucin-associated and tight junction proteins. In addition, it suppressed tumor growth by regulating apoptosis and proliferation. These results highlight the anti-tumor effects of CJ in an AOM/DSS-induced CAC mouse model.

Keywords: colorectal cancer; chronic inflammation; inflammatory bowel disease; cheonggukjang; anti-cancer effects; functional food

Citation: Lim, H.-J.; Park, I.-S.; Jeong, S.-J.; Ha, G.-S.; Yang, H.-J.; Jeong, D.-Y.; Kim, S.-Y.; Jung, C.-H. Effects of Cheonggukjang (Fermented Soybean) on the Development of Colitis-Associated Colorectal Cancer in Mice. *Foods* **2023**, *12*, 383. https://doi.org/10.3390/foods12020383

Academic Editors: Jennifer Ahn-Jarvis and Brittany A. Hazard

Received: 19 December 2022
Revised: 9 January 2023
Accepted: 11 January 2023
Published: 13 January 2023

1. Introduction

Colorectal cancer (CRC) is the third most common type of cancer in both men and women worldwide [1]. In 2018, the overall CRC incidence rate in Korea was 11.4%, 12.9% in men and 9.8% in women, with a mortality rate of 10% in men and 12.6% in women [2]. Chronic inflammation leads to various organ-specific diseases, depending on where the inflammation occurs [3]. In particular, chronic inflammation in the intestine causes inflammatory bowel disease (IBD), which increases the risk of developing CRC, also known as colitis-associated colorectal cancer (CAC) [4].

IBD is associated with multiple genetic, microbial, environmental, and immune-mediated factors [5,6]. However, the accurate etiology of IBD is still unknown. Recent studies have indicated that pro-inflammatory cytokines and gut dysbiosis induce both intestinal inflammation and the disruption of normal mucosal immunity, resulting in IBD [7,8]. Consequently, the suppression of intestinal inflammation may help prevent inflammation-associated colon cancer [9].

Cheonggukjang (CJ) is a soybean paste made through the fermentation of boiled soybeans. It is rich in substances that exhibit various biological activities, such as poly γ-glutamic acid (γ-PGA), isoflavone, saponins, phenolic acids, and flavonoids [10]. In many studies, CJ has emerged as a functional food with anti-obesity, antioxidant, and anti-inflammatory effects in bowel disease [10–12]. We have previously shown that CJ contains beneficial probiotics and is effective in preventing inflammatory diseases by suppressing the inflammatory signaling pathway in dextran sodium sulfate (DSS)-induced colitis mice [13]. However, to the best of our knowledge, the preventive effect of CJ on inflammatory colorectal tumorigenesis has not been elucidated. In this study, we report the anti-tumor growth effects of CJ by inhibiting azoxymethane (AOM)/DSS-induced pathological symptoms in an AOM/DSS-induced CAC mouse model.

2. Materials and Methods

2.1. Preparation of the CJ

The CJ was obtained from the Microbial Institute for Fermentation Industry (Sunchang-gun, Jeollabuk-do, Republic of Korea) as described previously [13]. It was produced using the traditional method of Kangjin-gun (Jeollanam-do, Republic of Korea) and had a moisture content of 53.15%. The CJ was dissolved in distilled water at 500 mg/kg and then stored at −20 °C before being orally administered to mice.

2.2. AOM/DSS-Induced Colorectal Cancer Model and CJ Treatment

A total of 32 male BALB/c mice (five-week-old) were purchased from Damool Science (Daejeon, Republic of Korea). The mice were maintained at 20–24 °C with a 12 h light/dark cycle and a relative humidity of 50–60%. After 1 week of acclimatization, the mice were divided into four groups (n = 8): NOR (normal group; only water), CON (control group; AOM (10 mg/kg)/DSS (2%)), PC (positive control group; AOM (10 mg/kg)/DSS (2%) and 5-aminosalicylic acid (75 mg/kg/day, 5-ASA, Sigma-Aldrich, St. Louis, MO, USA)), and CJ (CJ treated group; AOM (10 mg/kg)/DSS (2%) and CJ (100 mg/kg/day)). AOM/DSS-induced CAC was initiated by injecting the mice intraperitoneally with AOM (10 mg/kg, Sigma-Aldrich, St. Louis, MO, USA) on day 0. After 1 week, the mice were administered DSS (2%, MP Biomedicals, Irvine, CA, USA) for another week, followed by water for 2 weeks for recovery. This treatment was repeated twice (days 7–14, 32–39). All experimental animal procedures were performed within the guidelines of and approved by the Jeonju AgroBio-Materials Institute's Animal Care and Use Committee (JAMI IACUC 2022004).

2.3. ELISA Analysis

The serum levels of TNF-α (MTA00B), IFN-γ (MIF00), IL-6 (M6000B), IL-1β (MLB00C), IL-4 (M4000B), and IL-10 (M1000B) were analyzed using commercial ELISA kits (R&D Systems, Minneapolis, MN, USA) according to the manufacturer's recommendations.

2.4. Quantitative Real-Time PCR (qRT-PCR) Analysis

Total RNA was extracted from colonic tissues using a Hybrid-R™ kit (GeneAll, Seoul, Republic of Korea). First, cDNA was synthesized with the BioFACT™ 2X RT Pre-Mix (BIOFACT, Daejeon, Republic of Korea). Next, qRT-PCR was undertaken using the BioFACT™ 2X Real-Time PCR Master Mix (Bio-Fact) and analyzed with a sequence detection system (CFX96, Bio-Rad, Hercules, CA, USA). The primer sequences used are listed in Table 1.

Table 1. Primer sequences.

Gene	Forward (5′-3′)	Reverse (5′-3′)
TNF-α	CTGAACTTCGGGGTGATCGG	GGCTTGTCACTCGAATTTTGAGA
IFN-γ	AGCCCTATTACAGCACAG	TTCTAACAACAAGTATCCC
IL-6	TGTCTATACCACTTCACAAGTCGGAG	GCACAACTCTTTTCTCATTTCCAC
IL-1β	CAACCAACAAGTGATATTCTCCATG	GATCCACACTCTCCAGCTGCA
IL-4	GGTCTCAACCCCCAGCTAGT	GCCGATGATCTCTCTCAAGTGAT
IL-10	CTTACTGACTGGCATGAGGATCA	GCAGCTCTAGGAGCATGTGG
iNOS	CGAAACGCTTCACTTCCAA	TGAGCCTATATTGCTGTGGCT
COX-2	TTTGGTCTGGTGCCTGGTC	CTGCTGGTTTGGAATAGTTGCTC

Table 1. Cont.

Gene	Forward (5′-3′)	Reverse (5′-3′)
p53	CCCCTGTCATCTTTTGTCCCT	AGCTGGCAGAATAGCTTATTGAG
Bcl-2	GCTACCGTCGTGACTTCGC	CCCCACCGAACTCAAAGAAGG
Bcl-X$_L$	GGCACTGTGCGTGGAAAGCGTA	CCGCCGTTCTCCTGGATCCA
Bax	AGACAGGGGCCTTTTTGCTAC	AATTCGCCGGAGACACTCG
MUC-2	ATGCCCACCTCCTCAAAGAC	GTAGTTTCCGTTGGAACAGTGAA
Occludin	TCTGCTTCATCGCTTCCTTAG	GTCGGGTTCACTCCCATTA
ZO-1	AGGACACCAAAGCATGTGAG	GGCATTCCTGCTGGTTACA
GAPDH	GACGGCCGCATCTTCTTGT	CAGTGCCAGCCTCGTCCCGTACAA

2.5. Western Blotting

Western blotting was performed as described previously [13]. Briefly, about 20 μg of cell lysate was separated using a 10% SDS-PAGE and transferred to a PVDF membrane using the Trans-Blot Turbo Transfer system (Bio-Rad). The membranes were probed with specific antibodies: anti-iNOS, anti-COX-2, anti-p-p65, anti-p65, anti-Bax, anti-p53, anti-Bcl-2, anti-Bcl-X$_L$, and anti-β-actin (Cell Signaling Technology, Danvers, MA, USA). The protein–antibody binding was visualized using the enhanced chemiluminescence (ECL) detection system (Amersham Imager 600, GE Healthcare, Chicago, IL, USA).

2.6. Hematoxylin & Eosin (H&E) Staining and Immunohistochemistry

Colon tissue sections (4 μm thick) were stained with H&E. For immunolabeling, the sections were incubated with specific antibodies (anti-Muc2, anti-ZO-1, anti-occludin, anti-proliferating cell nuclear antigen (PCNA; Cell Signaling Technology, Danvers, MA, USA), and anti-Ki-67 (Abcam, Boston, MA, USA)) overnight at 4 °C in the dark. Next, the sections were incubated with either mouse or rabbit Envision plus polymer reagent (Dako, Glostrup Kommune, Denmark) for 30 min at 4 °C, and binding was detected by 3, 3′-diaminobenzidine (DAB) staining. Images were captured using a tissue slide scanner (Motic; Xiamen, China).

2.7. Statistical Analysis

A comparison of the statistical significance between groups was determined using a one-way ANOVA and a Tukey post-hoc test in GraphPad Prism (version 5.0; GraphPad Software, Inc., San Diego, CA, USA). The data are presented as mean ± standard deviation (SD). The statistical significance was set at $p < 0.05$.

3. Results and Discussion

3.1. CJ Attenuates Pathological Symptoms in Mice with AOM/DSS-Induced CAC

Chronic inflammation, one of the major contributing factors in the development of IBD, is closely associated with CRC and can be prevented or delayed with anti-inflammatory agents [14,15]. We have previously studied the anti-inflammatory effects of CJ [13] and now aimed to determine the anti-cancer effects of CJ in CRC, which is closely related to chronic inflammation. Therefore, in this study, we investigated the anti-cancer effects of CJ using an AOM/DSS-induced CAC mouse model (Figure 1a). The administration of a 2% DSS reduced the body weights of AOM/DSS-treated mice. However, the lost body weight was regained when the DSS administration was discontinued (Figure 1b). The decrease in body weight was moderate in the PC (5-ASA treated) and CJ groups. Based on previous reports of reduced colon length in mice with AOM/DSS-induced CAC [16], we measured the colon lengths of the treated mice to evaluate the protective effects of CJ. Compared with the NOR group, the AOM/DSS-treated group (CON) had significantly shorter colons. However, the administration of 5-ASA and CJ restored the colon length (Figure 1c,d). Next, we measured the spleen weights, an indicator of the severity of tumor progression, of AOM/DSS-induced CAC mice [16]. The CON group showed an increase in spleen weight, which was reversed by the administration of 5-ASA and CJ (Figure 1e), supporting the anti-tumor effects of CJ. These findings indicate that CJ attenuates the pathological symptoms in mice with AOM/DSS-induced CAC.

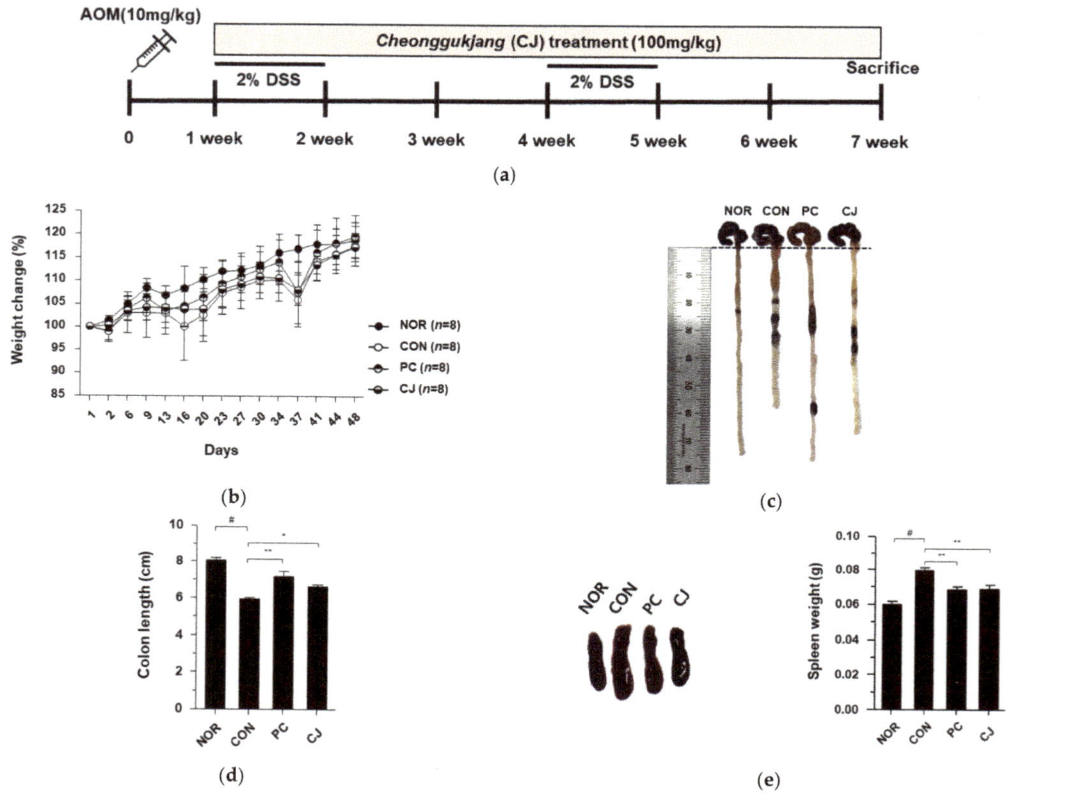

Figure 1. Effects of CJ on pathological symptoms of AOM/DSS-induced colorectal cancer (CAC) mice. (**a**) Experimental design; (**b**) weight change (%); (**c**) representative pictures of colon in mice; (**d**) colon length in mice; (**e**) representative pictures of spleen and spleen weight in mice. Values are means ± standard deviation (n = 8); NOR, normal group; CON, control group; PC, positive control group; CJ, cheonggukjang treated group; #, $p < 0.05$ versus normal group; **, $p < 0.005$; *, $p < 0.05$ versus control group.

3.2. *CJ Suppresses Tumorigenesis in AOM/DSS-Induced CAC Mouse Model*

AOM/DSS-induced CAC in mice is characterized by severe colitis with weight loss and a reduction in colon length, followed by the development of multiple colon tumors [17]. Therefore, we investigated the effects of CJ on intestinal tumorigenesis in mice with AOM/DSS-induced CAC. The AOM/DSS-treated mice showed an increase in macroscopic tumors, which was reversed following the administration of 5-ASA and CJ (Figure 2a.). The increased number of tumors in the CON group was reduced following the administration of 5-ASA and CJ (Figure 2b). Moreover, a significant difference was seen in the mean tumor size between the CON group and the PC and CJ groups (Figure 2c). Tumor formation and the thickening of the mucous membrane have been reported to increase the ratio of colon weight to colon length [18]. Therefore, we calculated the ratio of colon weight to colon length to confirm the inhibitory effects of CJ on tumorigenesis. As shown in Figure 2d, the weight-to-length ratio was significantly higher in the CON group than in the NOR group. However, this increase was significantly attenuated by the administration of 5-ASA and CJ

(Figure 2d). H&E staining is used for the histological diagnosis of various diseases, including IBD and cancer [19,20]. The pathological features of AOM/DSS-induced CAC in mice include inflammatory cell infiltration, crypt abscesses, and hyperchromatic nuclei [17,21]. The histopathological features related to CRC development were examined with H&E staining. Inflammatory cell infiltration and hyperchromatic nuclei were observed in the colon tissue of the CON group, and they were significantly ameliorated by the administration of PC and CJ (Figure 2e). These results indicate that CJ suppresses AOM/DSS-induced colitis-associated tumorigenesis in mice.

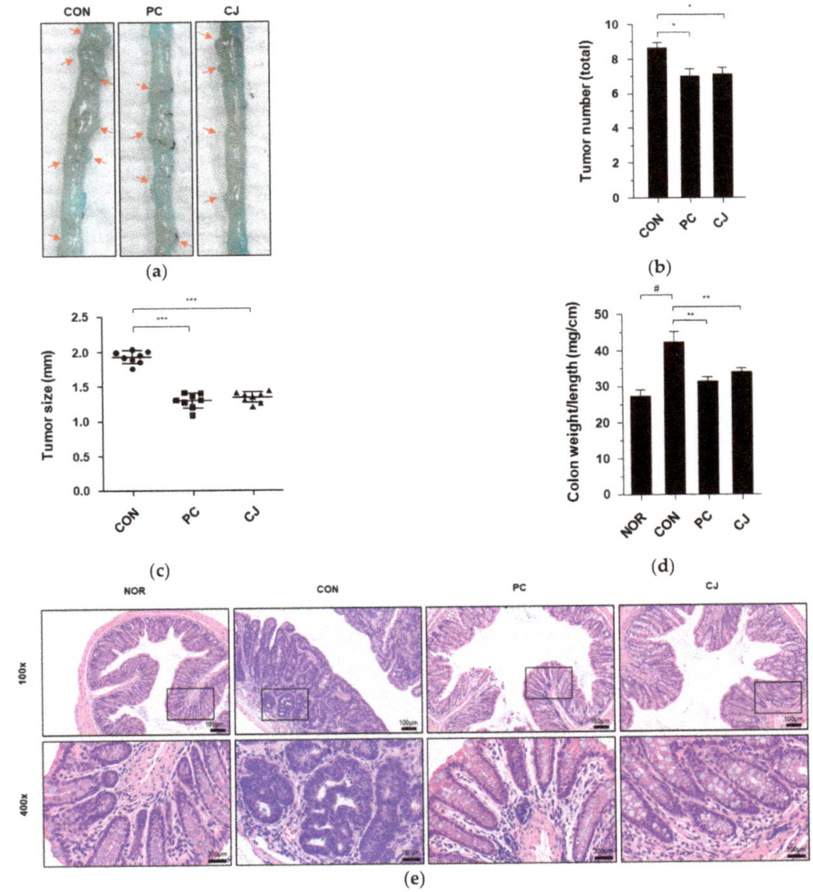

Figure 2. Effects of CJ on tumorigenesis in mice with AOM/DSS-induced CAC. (**a**) Representative pictures of colon tissue in the different groups of mice. Arrow indicates a tumor. Shown are the (**b**) total number of tumors, (**c**) tumor size, and (**d**) ratio of colon weight-to-colon length in the different groups of mice. All values are mean ± standard deviation ($n = 8$). (**e**) Representative histology for H&E staining of colon tissue. Magnification, 100× and 400×; Scale bar; 100 μm and 300 μm; NOR, normal group; CON, control group; PC, positive control group; CJ, cheonggukjang treated group. #, $p < 0.05$ versus normal group; ***, $p < 0.001$; **, $p < 0.005$; *, $p < 0.05$ versus control group.

3.3. CJ Modulates the Expression of Inflammatory Cytokines in AOM/DSS-Induced CAC Mice

AOM/DSS-induced CAC is associated with the expression of pro-inflammatory and anti-inflammatory cytokines, such as TNF-α, IFN-γ, IL-1β, IL-6, IL-4, and IL-10 [22,23]. As shown in Figure 3a-f, mRNA levels of pro-inflammatory cytokines (TNF-α, IFN-γ, IL-1β, and IL-6) were increased in the CON group, whereas those of anti-inflammatory cytokines (IL-4 and IL-10) were decreased. However, these changes in mRNA levels were reversed by the administration of 5-ASA and CJ (Figure 3a–f). To confirm these effects, we evaluated the levels of TNF-α, IFN-γ, IL-1β, IL-6, IL-4, and IL-10 in a serum by ELISA. Consistent with the mRNA results, the CON group showed increased TNF-α, IFN-γ, IL-1β, and IL-6 levels and decreased IL-4 and IL-10 levels (Figure 3g–l). The increase/decrease in cytokine levels was reversed in the PC and CJ groups, indicating that CJ modulates the expression of pro-inflammatory and anti-inflammatory cytokines in mice with AOM/DSS-induced CAC.

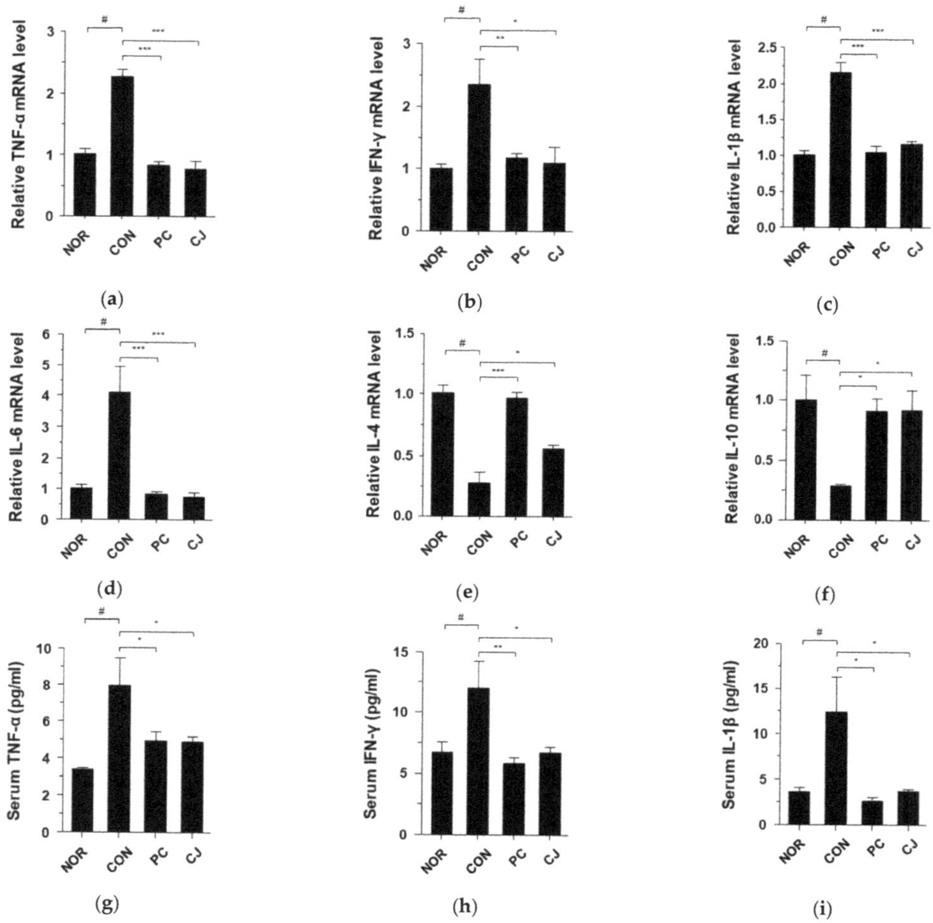

Figure 3. *Cont.*

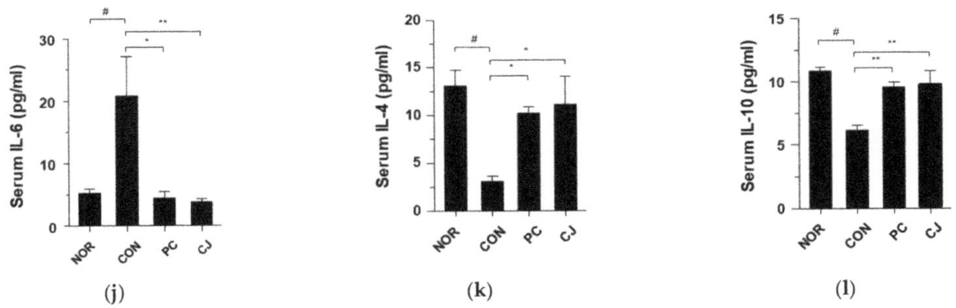

Figure 3. Effects of CJ on the expression of inflammatory cytokines in mice with AOM/DSS-induced CAC. Shown are the mRNA levels of (**a**) TNF-α, (**b**) IFN-γ, (**c**) IL-1β, (**d**) IL-6, (**e**) IL-4, and (**f**) IL-10 in the colon and protein levels of (**g**) TNF-α, (**h**) IFN-γ, (**i**), IL-1β, (**j**) IL-6, (**k**) IL-4, and (**l**) IL-10 in the serum. Values are mean ± standard deviation (n = 8); NOR, normal group; CON, control group; PC, positive control group; CJ, cheonggukjang treated group. #, $p < 0.05$ versus normal group; ***, $p < 0.001$; **, $p < 0.005$; *, $p < 0.05$ versus control group.

3.4. CJ Suppresses Activation of NF-κB Signaling in AOM/DSS-Induced CAC Mice

Nuclear factor-κB (NF-κB) is involved in inflammatory responses, proliferation, differentiation, and apoptosis. Moreover, the activation of NF-κB contributes to IBD and tumor development [24,25]. In a previous study, we showed that the NF-κB signaling pathway is activated by DSS and suppressed by CJ in DSS-treated mice [13]. Therefore, we investigated whether CJ suppresses NF-κB activation in mice with AOM/DSS-induced CAC. The CON group showed a significant increase in phospho-p65 NF-κB expression, which was decreased in the PC and CJ groups (Figure 4a). NF-κB activation induced several inflammatory enzymes, such as inducible nitric oxide synthase (iNOS) and cyclooxygenase (COX-2) [26]. Therefore, western blotting and quantitative real-time PCR (qRT-PCR) were performed to evaluate the iNOS and COX-2 levels in the treated mice. The protein levels of iNOS and COX-2 were increased in the CON group (Figure 4a). However, 5-ASA and CJ administration inhibited iNOS and COX-2 expression (Figure 4a). These findings were confirmed by the mRNA levels of iNOS and COX-2 measured by qRT-PCR (Figure 4b,c). These findings indicate that CJ suppresses the phospho-NF-κB, iNOS, and COX-2 expression in mice with AOM/DSS-induced CAC.

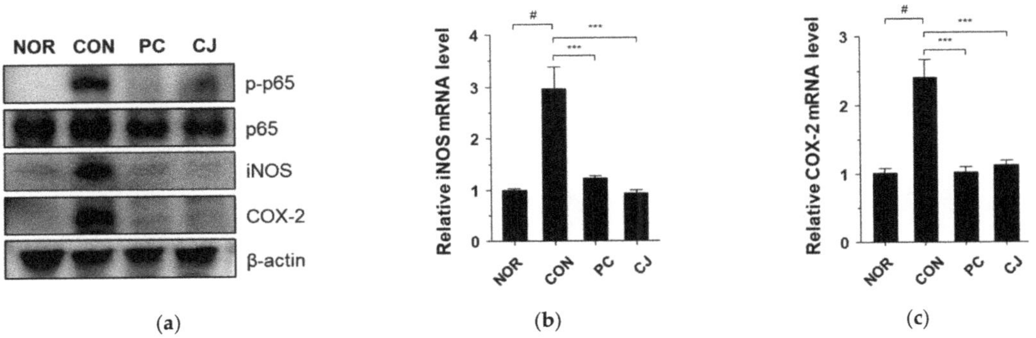

Figure 4. Effects of CJ on AOM/DSS-induced CAC-associated signaling pathways: (**a**) protein levels of phosphorylated p65, p65, iNOS, and COX2; and mRNA levels of (**b**) iNOS and (**c**) COX-2. Values are means ± standard deviation (n = 8); NOR, normal group; CON, control group; PC, positive control group; CJ, cheonggukjang treated group. #, $p < 0.05$ versus normal group; ***, $p < 0.001$ versus control group.

3.5. CJ Improves Intestinal Integrity in Mice with AOM/DSS-Induced CAC

The colonic mucus layer and tight junctions act as a layered defensive barrier in the colon and play a vital role in regulating mucosal permeability to ions, nutrients, and water [27,28]. The mucus layer, protected by the secreted mucin protein MUC2, is in contact with the epithelial cells lining the intestine and is joined via tight junctions [29]. Tight junctions consist of transmembrane proteins such as occludin, claudin, and junctional adhesion molecules [30]. Occludin directly interacts with zonula occludens-1 (ZO-1) and regulates paracellular permeability [31]. The loss of the colonic mucus layer is responsible for various intestinal diseases, including IBD and CRC [32,33].

We investigated the levels of mucin-associated protein (MUC2) and tight junction structural proteins (occludin and ZO-1) in the treated mice to assess the effects of AOM/DSS treatment on the colonic mucus layer. The mRNA levels of MUC2, occludin, and ZO-1 were significantly decreased in the CON group (Figure 5a–c). However, the administration of 5-ASA and CJ reversed the effects of AOM/DSS and significantly increased the mRNA levels of these markers (Figure 5a–c). Immunohistochemical staining confirmed the reduced expression of MUC2, Occludin, and ZO-1 in the CON group (Figure 5d). However, the administration of 5-ASA and CJ reversed these effects and increased the levels of MUC2, occludin, and ZO-1. These findings indicate that CJ may improve intestinal integrity by inducing the expression of mucin-associated and tight junction proteins in mice with AOM/DSS-induced CAC.

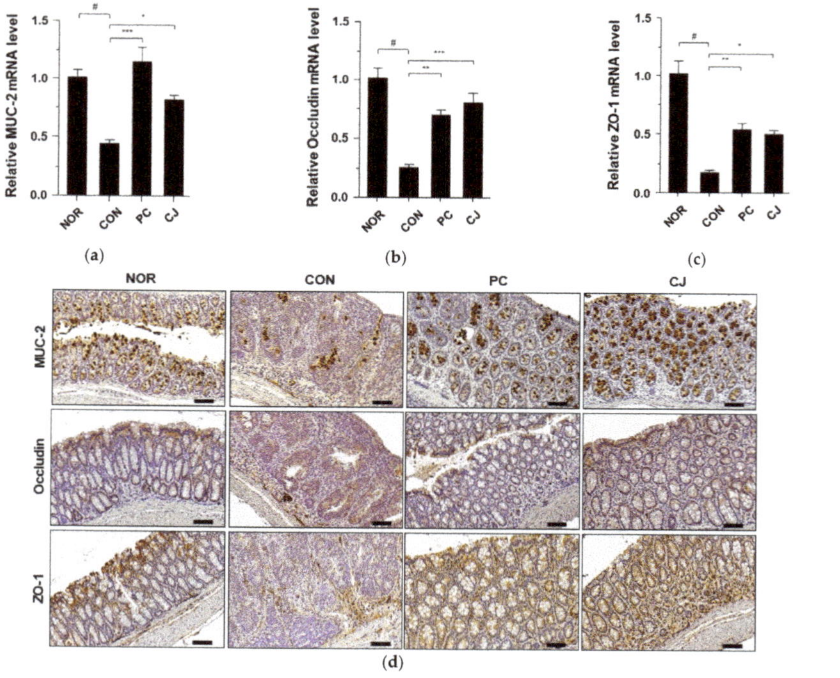

Figure 5. Effects of CJ on the expression of mucin-associated protein and tight junction proteins in mice with AOM/DSS-induced CAC. Shown are the mRNA levels of (**a**) Muc2, (**b**) occludin, and (**c**) ZO-1 in the different groups of mice. (**d**) Representative images of the colonic tissue immunohistochemically stained with MUC2, occludin, and ZO-1. Magnification, 200 X; scale bar, 60 μm. Values are means ± standard deviation (n = 8); NOR, normal group; CON, control group; PC, positive control group; CJ, cheonggukjang treated group. #, $p < 0.05$ versus normal group; ***, $p < 0.001$; **, $p < 0.005$; *, $p < 0.05$ versus control group.

3.6. CJ Suppresses Tumor Growth by Regulating Apoptosis and Proliferation

The occurrence of CRC is closely associated with abnormal cell growth via the rapid proliferation and evasion of apoptosis [34]. The NF-κB signaling pathway regulates apoptosis and cell proliferation factors such as p53, Bax, Bcl-2, Bcl-X_L, PCNA, and Ki67 [24,35]. In Figure 4a, we observed that the NF-κB signaling pathway was activated in the CON group. Thus, we evaluated the apoptosis and cell proliferation factors following treatment with AOM/DSS in this group. As shown in Figure 6a–d, the mRNA levels of pro-apoptotic markers, p53 and Bax, were significantly lower, and the mRNA levels of anti-apoptotic markers, Bcl-2 and Bcl-X_L, were higher in the AOM/DSS-treated CON group compared with the NOR group. Treatment with 5-ASA and CJ reversed these effects (Figure 6a–d). Western blotting confirmed these mRNA results at the protein level (Figure 6e). Next, to evaluate the regulation of cell proliferation by AOM/DSS, cell proliferation markers were studied with immunohistochemical staining. PCNA and Ki-67 expression were higher in the AOM/DSS-treated CON group compared with the NOR group, while 5-ASA and CJ reversed this effect (Figure 6f). These findings indicate that CJ inhibits tumor growth by inducing apoptosis and suppressing the proliferation of tumor cells in mice with AOM/DSS-induced CAC.

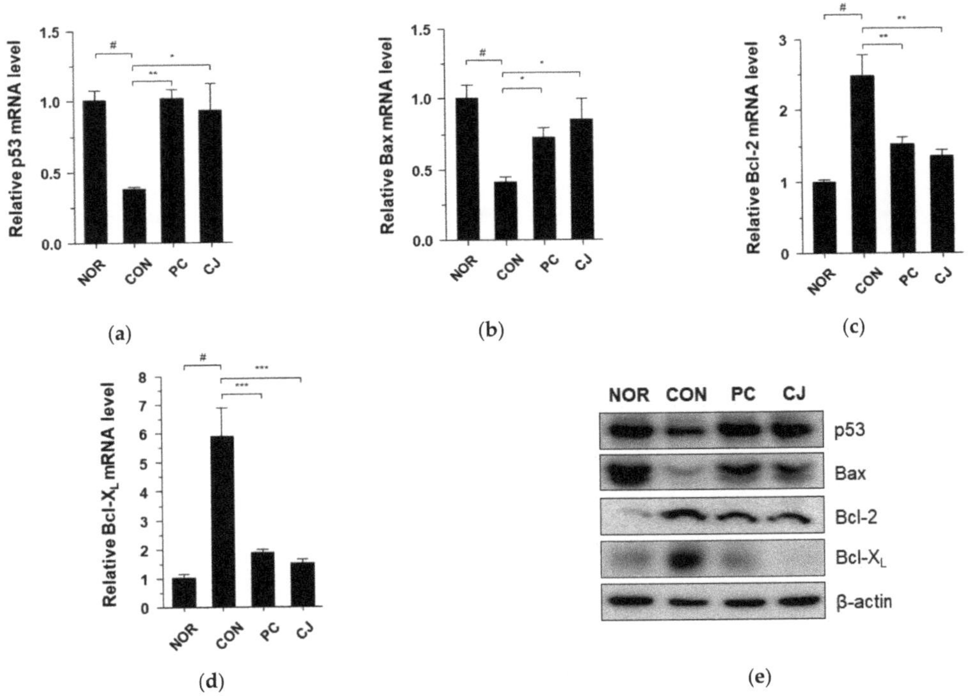

Figure 6. *Cont.*

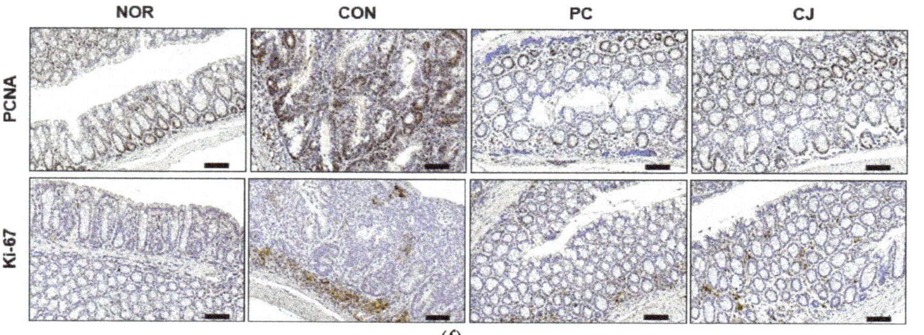

(f)

Figure 6. Effects of CJ on the expression of apoptosis and proliferation-related proteins in mice with DSS-induced colitis. mRNA levels of (**a**) p53, (**b**) Bax, (**c**) Bcl-2, and (**d**) Bcl-X_L in the different groups of mice were measured by qRT-PCR. (**e**) Protein levels of p53, Bax, Bcl-2, and Bcl-X_L were determined by western blotting. (**f**) Representative images of the colonic tissue immunohistochemically stained with PCNA and Ki67. Magnification, 200×; Scale bar, 60 μm. Values are means ± standard deviation (n = 8); NOR, normal group; CON, control group; PC, positive control group; CJ, cheonggukjang treated group. #, $p < 0.05$ versus normal group; ***, $p < 0.001$; **, $p < 0.005$; *, $p < 0.05$ versus control group.

4. Conclusions

This study demonstrates that CJ, made from the fermentation of boiled soybeans, significantly ameliorates AOM/DSS-induced pathological symptoms in mice including colonic shortening, spleen hypertrophy, tumor formation, and histopathological colonic changes, similar to the positive control (5-ASA). We also show that CJ and 5-ASA suppress the expression of pro-inflammatory cytokines and inflammatory mediators while increasing the expression of anti-inflammatory cytokines through the activation of NF-κB signaling pathways. Furthermore, CJ and 5-ASA attenuate tumorigenesis by inhibiting tumor growth through the induction of apoptosis and the inhibition of cell proliferation. These results suggest that CJ, which shows anti-cancer effects similar to 5-ASA, can be used as a functional food to prevent chronic inflammatory CRC.

Author Contributions: Conceptualization, S.-Y.K. and C.-H.J.; investigation, H.-J.L., I.-S.P., S.-J.J., G.-S.H. and H.-J.Y.; writing—original draft preparation, C.-H.J.; writing—review and editing, S.-Y.K. and C.-H.J.; funding acquisition, D.-Y.J. and S.-Y.K. All authors have read and agreed to the published version of the manuscript.

Funding: This research was funded by "Functional research of fermented soybean food (safety monitoring)" under the Ministry of Agriculture, Food, and Rural Affairs, and partly Korea Agro-Fisheries and Food Trade Corporation in 2022.

Institutional Review Board Statement: The animal study protocol was approved by the Animal Care Committee of Jeonju AgroBio-Materials Institute (JAMI IACUC 2022004, Jeonju, Republic of Korea).

Data Availability Statement: Data is contained within the article.

Conflicts of Interest: The authors declare no conflict of interest.

Abbreviation

CRC	Colorectal Cancer
IBD	Inflammatory Bowel Disease
CAC	Colitis-Associated Colorectal Cancer
CJ	Cheonggukjang
γ-PGA	poly γ-glutamic acid
DSS	Dextran Sulfate Sodium
AOM	Azoxymethane
NOR	Normal
CON	Control
PC	Positive Control
5-ASA	5-Aminosalicylic acid
qRT-PCR	Quantitative Real-Time PCR
H&E	Hematoxylin & Eosin
NF-κB	Nuclear Factor-κB
iNOS	Inducible Nitric Oxide Synthase
COX-2	Cyclooxygenase

References

1. Siegel, R.L.; Miller, K.D.; Goding Sauer, A.; Fedewa, S.A.; Butterly, L.F.; Anderson, J.C.; Cercek, A.; Smith, R.A.; Jemal, A. Colorectal cancer statistics, 2020. *CA Cancer J. Clin.* **2020**, *70*, 145–164. [CrossRef] [PubMed]
2. Hong, S.; Won, Y.J.; Lee, J.J.; Jung, K.W.; Kong, H.J.; Im, J.S.; Seo, H.G. Community of Population-Based Regional Cancer Registries. Cancer Statistics in Korea: Incidence, Mortality, Survival, and Prevalence in 2018. *Cancer Res. Treat.* **2021**, *53*, 301–315. [CrossRef] [PubMed]
3. Chen, L.; Deng, H.; Cui, H.; Fang, J.; Zuo, Z.; Deng, J.; Li, Y.; Wang, X.; Zhao, L. Inflammatory responses and inflammation-associated diseases in organs. *Oncotarget* **2017**, *9*, 7204–7218. [CrossRef] [PubMed]
4. Lasry, A.; Zinger, A.; Ben-Neriah, Y. Inflammatory networks underlying colorectal cancer. *Nat. Immunol.* **2016**, *17*, 230–240. [CrossRef] [PubMed]
5. Rieder, F.; Fiocchi, C.; Rogler, G. Mechanisms, Management, and Treatment of Fibrosis in Patients with Inflammatory Bowel Diseases. *Gastroenterology* **2017**, *152*, 340–350. [CrossRef] [PubMed]
6. Guan, Q. A Comprehensive Review and Update on the Pathogenesis of Inflammatory Bowel Disease. *J. Immunol. Res.* **2019**, *2019*, 7247238. [CrossRef] [PubMed]
7. Schetter, A.J.; Heegaard, N.H.; Harris, C.C. Inflammation and cancer: Interweaving microRNA, free radical, cytokine and p53 pathways. *Carcinogenesis* **2010**, *31*, 37–49. [CrossRef]
8. Ni, J.; Wu, G.D.; Albenberg, L.; Tomov, V.T. Gut microbiota and IBD: Causation or correlation? *Nat. Rev. Gastroenterol. Hepatol.* **2017**, *14*, 573–584. [CrossRef]
9. Teng, S.; Hao, J.; Bi, H.; Li, C.; Zhang, Y.; Zhang, Y.; Han, W.; Wang, D. The Protection of Crocin Against Ulcerative Colitis and Colorectal Cancer via Suppression of NF-κB-Mediated Inflammation. *Front. Pharmacol.* **2021**, *12*, 639458. [CrossRef] [PubMed]
10. Kim, I.S.; Hwang, C.W.; Yang, W.S.; Kim, C.H. Current Perspectives on the Physiological Activities of Fermented Soybean-Derived Cheonggukjang. *Int. J. Mol. Sci.* **2021**, *22*, 5746. [CrossRef]
11. Jang, C.H.; Oh, J.; Lim, J.S.; Kim, H.J.; Kim, J.S. Fermented Soy Products: Beneficial Potential in Neurodegenerative Diseases. *Foods* **2021**, *10*, 636. [CrossRef]
12. Jeong, D.Y.; Ryu, M.S.; Yang, H.J.; Park, S. γ-PGA-Rich Chungkookjang, Short-Term Fermented Soybeans: Prevents Memory Impairment by Modulating Brain Insulin Sensitivity, Neuro-Inflammation, and the Gut-Microbiome-Brain Axis. *Foods* **2021**, *10*, 221. [CrossRef]
13. Lim, H.J.; Kim, H.R.; Jeong, S.J.; Yang, H.J.; Ryu, M.S.; Jeong, D.Y.; Kim, S.Y.; Jung, C.H. Protective Effects of Fermented Soybeans (Cheonggukjang) on Dextran Sodium Sulfate (DSS)-Induced Colitis in a Mouse Model. *Foods* **2022**, *11*, 776. [CrossRef] [PubMed]
14. Lucafò, M.; Curci, D.; Franzin, M.; Decorti, G.; Stocco, G. Inflammatory Bowel Disease and Risk of Colorectal Cancer: An Overview from Pathophysiology to Pharmacological Prevention. *Front. Pharmacol.* **2021**, *12*, 772101. [CrossRef] [PubMed]
15. van Staa, T.P.; Card, T.; Logan, R.F.; Leufkens, H.G. 5-Aminosalicylate use and colorectal cancer risk in inflammatory bowel disease: A large epidemiological study. *Gut* **2005**, *54*, 1573–1578. [CrossRef] [PubMed]
16. Jiang, W.; Li, Y.; Zhang, S.; Kong, G.; Li, Z. Association between cellular immune response and spleen weight in mice with hepatocellular carcinoma. *Oncol. Lett.* **2021**, *22*, 625. [CrossRef]
17. Parang, B.; Barrett, C.W.; Williams, C.S. AOM/DSS Model of Colitis-Associated Cancer. *Methods Mol. Biol.* **2016**, *1422*, 297–307.
18. Chen, L.H.; Song, J.L.; Qian, Y.; Zhao, X.; Suo, H.Y.; Li, J. Increased preventive effect on colon carcinogenesis by use of resistant starch (RS3) as the carrier for polysaccharide of Larimichthys crocea swimming bladder. *Int. J. Mol. Sci.* **2014**, *15*, 817–829. [CrossRef]
19. Cornaggia, M.; Leutner, M.; Mescoli, C.; Sturniolo, G.C.; Gullotta, R. Chronic idiopathic inflammatory bowel diseases: The histology report. *Dig. Liver Dis.* **2011**, *43*, S293–S303. [CrossRef]

20. Li, Y.; Li, N.; Yu, X.; Huang, K.; Zheng, T.; Cheng, X.; Zeng, S.; Liu, X. Hematoxylin and eosin staining of intact tissues via delipidation and ultrasound. *Sci. Rep.* **2018**, *8*, 12259. [CrossRef]
21. Wu, S.; Luo, W.; Wu, X.; Shen, Z.; Wang, X. Functional Phenotypes of Peritoneal Macrophages Upon AMD3100 Treatment during Colitis-Associated Tumorigenesis. *Front. Med.* **2022**, *9*, 840704. [CrossRef]
22. Oh, N.S.; Lee, J.Y.; Kim, Y.T.; Kim, S.H.; Lee, J.H. Cancer-protective effect of a synbiotic combination between Lactobacillus gasseri 505 and a Cudrania tricuspidata leaf extract on colitis-associated colorectal cancer. *Gut Microbes.* **2020**, *12*, 1785803. [CrossRef]
23. Zhu, J.; Paul, W.E. CD4 T cells: Fates, functions, and faults. *Blood* **2008**, *112*, 1557–1569. [CrossRef] [PubMed]
24. Brantley, D.M.; Chen, C.L.; Muraoka, R.S.; Bushdid, P.B.; Bradberry, J.L.; Kittrell, F.; Medina, D.; Matrisian, L.M.; Kerr, L.D.; Yull, F.E. Nuclear factor-kappaB (NF-kappaB) regulates proliferation and branching in mouse mammary epithelium. *Mol Biol Cell.* **2001**, *12*, 1445–1455. [CrossRef] [PubMed]
25. Oeckinghaus, A.; Ghosh, S. The NF-kappaB family of transcription factors and its regulation. *Cold Spring Harb. Perspect. Biol.* **2009**, *1*, a000034. [CrossRef] [PubMed]
26. Slattery, M.L.; Mullany, L.E.; Sakoda, L.; Samowitz, W.S.; Wolff, R.K.; Stevens, J.R.; Herrick, J.S. The NF-κB signalling pathway in colorectal cancer: Associations between dysregulated gene and miRNA expression. *J. Cancer Res. Clin. Oncol.* **2018**, *144*, 269–283. [CrossRef]
27. Grondin, J.A.; Kwon, Y.H.; Far, P.M.; Haq, S.; Khan, W.I. Mucins in Intestinal Mucosal Defense and Inflammation: Learning from Clinical and Experimental Studies. *Front. Immunol.* **2020**, *11*, 2054. [CrossRef]
28. Herath, M.; Hosie, S.; Bornstein, J.C.; Franks, A.E.; Hill-Yardin, E.L. The Role of the Gastrointestinal Mucus System in Intestinal Homeostasis: Implications for Neurological Disorders. *Front. Cell Infect. Microbiol.* **2020**, *10*, 248. [CrossRef] [PubMed]
29. Capaldo, C.T.; Powell, D.N.; Kalman, D. Layered defense: How mucus and tight junctions seal the intestinal barrier. *J. Mol. Med.* **2017**, *95*, 927–934. [CrossRef]
30. Zihni, C.; Mills, C.; Matter, K.; Balda, M.S. Tight junctions: From simple barriers to multifunctional molecular gates. *Nat. Rev. Mol. Cell Biol.* **2016**, *17*, 564–580. [CrossRef]
31. Odenwald, M.A.; Choi, W.; Buckley, A.; Shashikanth, N.; Joseph, N.E.; Wang, Y.; Warren, M.H.; Buschmann, M.M.; Pavlyuk, R.; Hildebrand, J.; et al. ZO-1 interactions with F-actin and occludin direct epithelial polarization and single lumen specification in 3D culture. *J. Cell Sci.* **2017**, *130*, 243–259. [CrossRef] [PubMed]
32. de Souza, H.S.; Fiocchi, C. Immunopathogenesis of IBD: Current state of the art. *Nat. Rev. Gastroenterol. Hepatol.* **2016**, *13*, 13–27. [CrossRef] [PubMed]
33. Clay, S.L.; Fonseca-Pereira, D.; Garrett, W.S. Colorectal cancer: The facts in the case of the microbiota. *J. Clin. Investig.* **2022**, *132*, e155101. [CrossRef]
34. Testa, U.; Pelosi, E.; Castelli, G. Colorectal cancer: Genetic abnormalities, tumor progression, tumor heterogeneity, clonal evolution and tumor-initiating cells. *Med. Sci.* **2018**, *6*, 31. [CrossRef] [PubMed]
35. Cui, X.; Shen, D.; Kong, C.; Zhang, Z.; Zeng, Y.; Lin, X.; Liu, X. NF-κB suppresses apoptosis and promotes bladder cancer cell proliferation by upregulating survivin expression in vitro and in vivo. *Sci. Rep.* **2017**, *7*, 40723. [CrossRef] [PubMed]

foods

MDPI

Article

Short-Term Soy Bread Intervention Leads to a Dose-Response Increase in Urinary Isoflavone Metabolites and Satiety in Chronic Pancreatitis

Jennifer H. Ahn-Jarvis [1,†], Daniel Sosh [1], Erin Lombardo [2,‡], Gregory B. Lesinski [3,§], Darwin L. Conwell [3,‖], Phil A. Hart [3] and Yael Vodovotz [1,*]

1 College of Food, Agricultural, and Environmental Sciences, Department of Food Science and Technology, The Ohio State University, Columbus, OH 43210, USA
2 College of Public Health, The Ohio State University, Columbus, OH 43210, USA
3 Division of Gastroenterology, Hepatology and Nutrition, The Ohio State University Wexner Medical Center, Columbus, OH 43210, USA
* Correspondence: vodovotz.1@osu.edu
† Current address: Food Innovation and Health, Quadram Institute Bioscience, Norwich, NR4 7UQ, UK.
‡ Current address: Partnership for Proactive Health, Inc., Charleston, SC 29455, USA.
§ Current address: Department of Hematology and Medical Oncology, School of Medicine, Emory University, Atlanta, GA 30322, USA.
‖ Current address: Internal Medicine, University of Kentucky College of Medicine, Lexington, KY 40536, USA.

Abstract: Patients with chronic pancreatitis (CP) are particularly vulnerable to nutrient malabsorption and undernutrition caused by the underlying pathology of their disease. Dietary intervention trials involving soy isoflavones in patients with CP are limited and isoflavone metabolites have not yet been reported. We hypothesized soy bread containing plant-based protein, dietary fiber, and isoflavones would be well-tolerated and restore gut functional capacity which would lead to isoflavone metabolites profiles like those of healthy populations. Participants (n = 9) received 1 week of soy bread in a dose-escalation design (1 to 3 slices/day) or a 4-week maximally tolerated dose (n = 1). Dietary adherence, satiety, and palatability were measured. Isoflavone metabolites from 24 h urine collections were quantified using high-performance liquid chromatography. A maximum dose of three slices (99 mg of isoflavones) of soy bread per day was achieved. Short-term exposure to soy bread showed a significant dose-response increase (p = 0.007) of total isoflavones and their metabolites in urine. With increasing slices of soy bread, dietary animal protein intake (p = 0.009) and perceived thirst ($p < 0.001$) significantly decreased with prolonged satiety ($p < 0.001$). In this study, adherence to short-term intervention with soy bread in CP patients was excellent. Soy isoflavones were reliably delivered. These findings provide the foundation for evaluating a well-characterized soy bread in supporting healthy nutrition and gut function in CP.

Keywords: soy bread; isoflavones; satiety; metabolites; chronic pancreatitis

Citation: Ahn-Jarvis, J.H.; Sosh, D.; Lombardo, E.; Lesinski, G.B.; Conwell, D.L.; Hart, P.A.; Vodovotz, Y. Short-Term Soy Bread Intervention Leads to a Dose-Response Increase in Urinary Isoflavone Metabolites and Satiety in Chronic Pancreatitis. *Foods* **2023**, *12*, 1762. https://doi.org/10.3390/foods12091762

Academic Editor: Cornelia Witthöft

Received: 1 March 2023
Revised: 19 April 2023
Accepted: 22 April 2023
Published: 24 April 2023

1. Introduction

Chronic pancreatitis (CP) is a syndrome characterized by chronic inflammation and fibrosis of the pancreas, and is commonly associated with debilitating effects on health outcomes and quality of life [1]. Etiologies include toxin exposures such as alcohol and smoking, recurrent acute pancreatitis, genetics, and idiopathic mechanisms [1]. Chronic inflammation and fibrosis lead to destruction of the parenchyma and progressive loss of endocrine and exocrine function [1]. Consequently, patients often suffer from abdominal pain, diabetes mellitus, and maldigestion. Unfortunately, there are currently no approved medical therapies to interrupt or reverse this process, and management primarily consists of screening for and treating complications [2]. Due to the underlying pathophysiology and functional outcomes of CP, patients are often at risk for malnutrition. Steatorrhea, malabsorption, alcoholism (in alcoholic CP), and poor dietary intake due to abdominal pain and other gastrointestinal symptoms frequently lead to nutrient deficiencies [3,4].

The culmination of maldigestion and malabsorption of nutrients due to exocrine and endocrine insufficiency leads to metabolic perturbations [5], dysbiosis [6–8], impaired immunity [9], and loss of intestinal functional capacity [10–12] in CP patients. Several studies have reported changes in intestinal microbiota in CP. For instance, a reduction in beneficial bacteria *Faecalibacterium prausnitzii* and *Ruminococcus bromii* was observed in CP subjects compared to healthy controls [7,10,11]. Both bacteria are responsible for production of butyrate, which has a critical role in the growth and differentiation of intestinal epithelial cells as well as supporting healthy immune and intestinal barrier function [13].

Current consensus for the nutritional management of malnutrition in CP patients has been focused on caloric support using high protein, high-energy foods; however, there has been very little attention paid towards recovery of intestinal functional capacity [14]. Soy is rich in proteins (~50% in defatted soy flour) and prebiotics (dietary fiber, linoleic acid, and isoflavones) [15,16] which work synergistically to restore intestinal mucosal integrity, reduce inflammation, improve microbiota composition, and enhance localized production of short chain fatty acids [17–20]. Specifically, when soy is delivered as a bread versus a beverage, the physicochemical properties of bread prolong the gut transit time and promote the formation of isoflavone metabolites [21,22]. Equol, a mammalian derived metabolite of daidzein, has been purported to display greater bioactivity than its parent compound [23] and it has been associated as a marker of health [24,25]. In previous studies, the frequency of equol-producing phenotypes with those following a Westernized diet [26] was ~30% whereas the frequency doubles to 60% in those consuming an Asian diet [27]. Bacteria belonging to the *Coriobacteriaceae* and *Clostridiaceae* families have been associated with equol-producing phenotypes [24,25,28]. However, the mere presence of these equol-producing bacteria did not ensure equol production; rather, it was a combination of a healthy lifestyle, high microbial diversity, and equol-producing bacteria [29].

Given that metabolic impairments and dysbiosis are commonly found in CP patients, alterations in isoflavone absorption and metabolism are possible. In vitro and animal studies have shown that isoflavones can restore gut functional capacity in gastrointestinal disorders [30–32]. Descriptive epidemiology, laboratory animal models, and in vitro studies provide evidence of the health benefits of soy consumption in chronic inflammatory disease; however, data from clinical intervention trials involving whole soy foods in patients with chronic pancreatitis (CP) are almost non-existent. Among the limited number of human studies evaluating the impact of isoflavones on gastrointestinal disease, none have reported isoflavone metabolites in CP [33–35]. Therefore, we examined isoflavone absorption and metabolism and dietary habits in an open label trial using soy bread administered for 1 or 4 weeks in subjects with CP. We previously reported that soy bread intervention was safe, associated with a decrease in the pro-inflammatory cytokine tumor necrosis factor alpha (TNFα), and possibly reduced other pro-inflammatory cytokines, such as interferon gamma and interleukin-6 [36]. CP is characterized by a state of chronic inflammation and increased levels of circulating pro-inflammatory cytokines; blunting this process is one purposed mechanism to intervene in the progression of CP [2,9]. Soy bread is a whole-food approach which allows for delivery of high-quality proteins, dietary fiber, and a complex mixture of soy bioactives to target inflammation and nutrients to offset malnutrition in patients with CP.

The current study was a pre-planned secondary analysis assessing the impact of soy bread intervention on isoflavone metabolism and dietary habits in CP participants during dose escalating (DE) and a maximally tolerated dose (MTD) phase. These analyses are needed to determine the potential adherence to soy bread intervention in future participants as well as the potential for inter-subject variability. We hypothesized soy bread containing high-quality plant-based protein, dietary fiber, and isoflavones would be well-tolerated and restore gut functional capacity, which would be evidenced by the presence of isoflavone metabolites like those of a healthy population. The primary objectives of this study were

(a) to evaluate tolerability of soy bread intervention in the context of palatability and satiety, and (b) to assess urinary isoflavone absorption and metabolism in CP. Overall, this small clinical trial will provide the necessary clinical data required to move this intervention forward in the setting of a future placebo-controlled trial for individuals with CP. In addition, the findings will provide valuable mechanistic information to understand inflammatory biomarkers relevant to CP and to facilitate in designing a future study of individual differences in soy isoflavone metabolism within this patient population.

2. Materials and Methods

2.1. Soy Bread Materials

A wheat bread where 30% of the wheat flour was displaced by a mixture of soymilk and soy flour was designed to deliver ~30 mg aglycone equivalents (AE) of total isoflavones/slice. Granulated sugar, kosher salt (Diamond Crystal, Savannah, GA, USA), vegetable shortening (Crisco; Orrville, OH, USA), bread flour (Bouncer; Bay State Milling, Minneapolis, MN, USA), and baker's yeast (Saf-Instant Yeast; Lesaffre, Marcq-en-Barœul, FR, USA) were purchased from Gordon Food Service (Dublin, OH, USA). Vital wheat gluten (Manildra Milling, Leawood, KS, USA) and defatted soy flour (Baker's Nutrisoy, Archer Daniel) were purchased from Skidmore Sales (Cincinnati, OH, USA). Soymilk powder (Benesoy; Devansoy, Carroll, IA, USA) was purchased from Devansoy. The standardized soymilk and soy flour mixture was produced using soymilk powder (25 kg) and soy flour (75 kg) that was blended into a single mixture using a large capacity ribbon blender (model 1608, Federal Equipment Co., Cleveland, OH, USA). The soy mix was packaged into 5 kg vacuum sealed bags and stored at −40 °C until used for bread preparation.

2.2. Soy Bread Preparation and Isoflavone Analysis

Soy breads were prepared as previously described using a sponge-dough process [37]. Sixty loaves (1300 g each) of bread from 6 batches were manufactured at The Ohio State University Wilbur Gould Food Industries Center over a two-week period. Finished breads were baked until an internal temperature of 96.0 ± 0.5 °C was reached, then cooled at ambient conditions for 4 h, and uniformly sliced (SM302B Doyon, Menominee, MI, USA). Individual slices were weighed and vacuum-sealed in polyethylene bags, then immediately frozen and stored (−40 °C) to extend bread quality and preserve isoflavone content. Frozen breads were distributed to patients in insulated bags supplied with ice packs to keep bread frozen for 2 h. Frozen loaves of bread were distributed with thawing instructions. Isoflavone composition of breads from each batch (n = 6) were analyzed using HPLC equipped with photodiode array detector (PDA) (Waters Corp, Milford, MA, USA). The HPLC equipment and conditions as well as preparation of the isoflavone standards are detailed by Ahn-Jarvis et al. [38].

2.3. Study Design

The design has been described in detail in our previous publication [36]. In brief, this feasibility study involved two sequential phases: In Phase 1, dose escalation (DE) was used to determine the highest tolerable dose of soy bread. In Phase 2, maximally tolerated dose (MTD) was used to assess tolerance of the maximum target dose during a 4-week intervention (Figure 1). Study subjects were allowed to participate at multiple dose levels.

2.4. Participant Description

The study protocol was approved by The Ohio State University Biomedical Science Institutional Review Board (1 October 2014) as protocol 2014H0226 and registered on clinicatrials.gov (NCT 02577640) (https://clinicaltrials.gov/ct2/show/NCT02577640?term=NCT+02577640&draw=2&rank=1 (accessed 23 April 2020). Participants with CP (>18 year of age) who met the eligibility criteria were evaluated, and informed consent was obtained at The Ohio State University Wexner Medical Center Pancreas Clinics. Participants were asked to abstain from legumes (legume-free diet), supplement their diet with the prescribed number of slices, and document the time of soy bread consumption in their daily journal. Condiments were permitted on the bread, but heating was discouraged since the latter may degrade the isoflavones. All participants were provided a standardized multivitamin/mineral supplement (One Daily Multiple Plus Minerals, CVS Caremark, Woonsocket, RI, USA) and asked to discontinue all other dietary supplements (vitamins and botanical supplements) for the duration of the study. Participants were instructed to document gastrointestinal symptoms, frequency of bowel movements, and score their stool using the Bristol stool scale. A registered dietitian provided instructions to complete 3-day diet records (days 2, 4, and 6 for DE and MTD phases as well as additional days −5, −3, and −1 during the MTD phase), questionnaires, and a 24 h urine collection. Fasting blood and spot urine were collected at the start of intervention (baseline) and at the end of the 7-day intervention for DE and every week for the MTD participant.

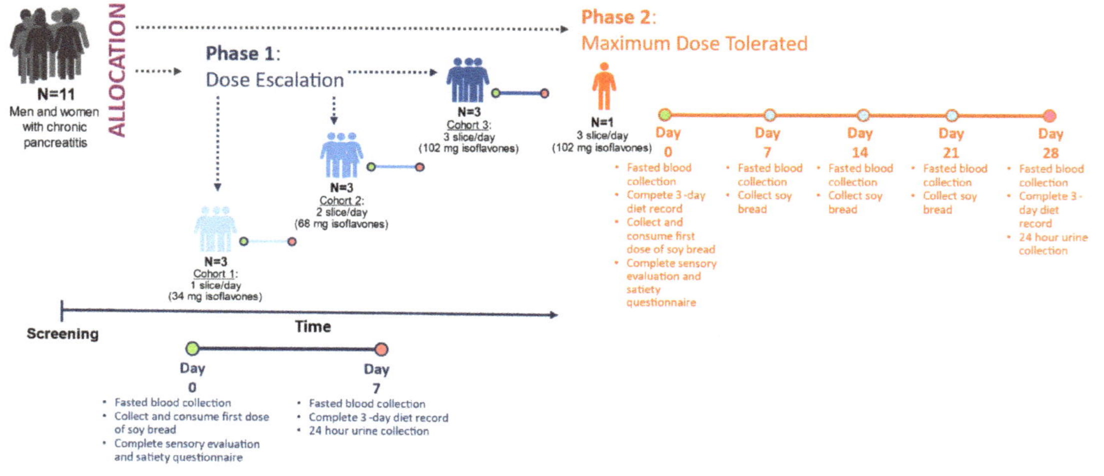

Figure 1. Study design of 3 + 3 dose escalating phase leading to maximum tolerated dose phase. Depending on the study phase, isoflavones were measured in urine on day 7 or day 28. Circles represent study visit days.

2.5. Sensory and Satiety Assessments

Participants were provided with a sensory ballot with written instructions to complete at home. They were to be completed on the first day, immediately after bread was consumed. Participants were instructed to rinse their mouths with cold tap water prior to eating their bread. A 9-point hedonic scale (1 = dislike extremely to 9 = like extremely) was used to assess six attributes (overall liking, aroma, flavor, sweetness, bitterness, and texture) of the soy bread and was followed by a descriptive analysis. It was performed using a horizontal line scale (6 inches) to evaluate whether soy bread attributes were like their ideal bread. Intensity was measured (1 = low and 10 = high) for brown crust color, sweetness, saltiness, bitterness, hardness, chewiness, and oiliness. Participants were given the following written lexicons to calibrate their senses to the respective attribute: peanut (1) and milk chocolate (10) for brown color; potato (1) and carrot (10) for sweetness; tomato (1) and olives (10) for saltiness; spinach (1) and coffee (10) for bitterness; cooked pasta (1) and licorice candy (10) for hardness; cream cheese (1) and raisin (10) for chewiness; nuts (1) and French fries (10) for oiliness. Additionally, measures of satiety (perceived fullness or lack of hunger) were conducted using a vertical (6 inches) visual analogue scale (VAS) to examine the impact of satiety on soy bread compliance. Subjects were asked to rate their satiety (hunger, fullness, how much they can eat, thought of food, desire, and urge to eat) as well rate their nausea, thirst, and energy. VAS was to be completed prior to the first dose of soy bread consumed, then 30 min and 4 h after consumption of the soy bread, as well as before going to bed. For the first day, participants were asked to consume a full dose as a single meal.

2.6. Laboratory and Clinical Measures of Toxicity

Fasting blood (electrolytes, glucose, renal and liver function tests) was collected at the enrollment and end of study visits. All clinical blood samples were analyzed by The Ohio State University Wexner Medical Center clinical laboratories.

2.7. Isoflavone Quantification in Urine

2.7.1. Chemicals

From Fisher Scientific (Fairlawn, OH, USA), anhydrous sodium acetate and solvents (diethyl ether, formic acid, acetonitrile, methanol, and water) which were HPLC-grade or higher were purchased. Lyophilized glucuronidase/sulfatase from Helix pomatia, dimethyl sulfoxide, and equol were obtained from Sigma-Aldrich (St. Louis, MO, USA). Daidzein, genistein, glycitein, daidzin, genistin, glycitin, acetyl daidzin, malonyl genistin, and acetyl genistin were obtained from LC laboratories division (PKC Pharmaceuticals, Inc., Woburn, MA, USA), and malonyl daidzin and malonyl glycitin from Wako Chemicals USA, Inc., (Richmond, VA, USA). Isoflavone metabolites (dihydrodaidzein DHD, dihydrogenistein DHG, ODMA, and 6-OH-ODMA) were acquired from Plantech (Reading, UK). The following authenticated isoflavone standards were purchased from Sigma-Aldrich (St. Louis, MO, USA): daidzein, genistein, glycitein, and equol. Food isoflavone standards (daidzin, genistin, glycitin, acetyl daidzin, malonyl genistin, and acetyl genistin) were obtained from LC Laboratories division (PKC Pharmaceuticals, Inc., Woburn, MA, USA).

2.7.2. Sample Handling and Isoflavone Quantification

Isoflavones were quantified using high-performance liquid chromatography (HPLC) grade chemicals and authenticated isoflavone standards as described by Ahn-Jarvis et al. 2015 [37]. Subjects were provided 24 h urine collection vessels, which were pre-weighed with 0.5 g/L boric acid, and collection began one day prior to completion of study (i.e., Day 6 and/or Day 27). Mass of urine combined with urine specific gravity was used to determine urine volume over 24 h. A minimum collection of 20 h was considered complete and used for analysis. Urine aliquots were stored at −80 °C for HPLC analysis. Extraction and quantification of isoflavonoids in urine are detailed by Ahn-Jarvis et al. [37]. Regression analysis of calibration curves revealed a good linear relationship (R^2 = 0.9996 ± 0.0002). The inter-assay and intra-assay variability in urine was less than 15%. Isoflavone concentrations in urine less than 43 nmol/L (12.2 ng/mL) to 54 nmol/L (13.7 ng/mL) for the parent (daidzein, genistein, glycitein) isoflavonoids and less than 48 nmol/L (13.1 ng/mL) to 62 nmol/L (15.9 ng/mL) for intermediate (DHG and DHD) isoflavonoids were considered below the level of quantification. Microbial metabolites of isoflavones less than 120 nmol/L (29.1 ng/mL) for equol, 88 nmol/L (22.7 ng/mL) for O-DMA, and 137 nmol/L (39.5 ng/mL) for 6 OH-ODMA were considered below the level of quantification.

2.8. Statistical Analysis

SPSS software (version 22, IBM, Armonk, NY, USA) was used for analyses. A paired t-test was used to evaluate the differences between baseline and at the end of study in clinical biomarkers. An ANOVA was used to discriminate differences in satiety parameters, isoflavone concentrations in urine, and dietary intake in respects to dose (slices/day) and/or time. When statistically significant ($p < 0.05$) differences were found, a Tukey's post hoc test was used to determine the mean separation. Hierarchical cluster analysis for urine isoflavones was conducted using proportions of metabolites belonging to the daidzein family as three components (equol, ODMA, and DHD + Daidzein). The clusters were formed using Euclidean distances and average linkage. These methods are detailed in our previous studies [37,39]. Sensory data were analyzed using Wilcoxon signed-rank test and reported as median ± inter-quartile range. Radar analysis compared soy bread attributes to participants' self-described ideal bread using paired *t*-test.

3. Results and Discussion

3.1. Soy Bread Nutrient and Isoflavone Composition

Each slice of soy bread provided ~34 mg total isoflavones. The nutrient and isoflavone composition of soy bread are detailed in Table 1. Soy bread isoflavone composition remained stable during 2 years of frozen storage.

Table 1. Nutrient and isoflavone content of one slice of soy bread.

Nutrition Parameters	
Serving Size (g)	92.4 ± 2.3
Nutrient Composition [1]	
Energy (kcal)	220.8
Fat (g)	3.4
Total Carbohydrates (g)	37.8
Dietary Fiber (g)	3.7
Protein (g)	15.2
Total Isoflavones (Mean ± SD, mg) [2]	**34.38 ± 0.21**
Daidzein (mg)	1.44 ± 0.10
Daidzin (mg)	3.81 ± 0.23
6″-O-Acetyldaidzin (mg)	0.54 ± 0.03
6″-O-Malonyldaidzin (mg)	6.48 ± 0.39
Genistein (mg)	1.84 ± 0.14
Genistin (mg)	6.92 ± 0.40
6″-O-Acetylgenistin (mg)	0.67 ± 0.03
6″-O-Malonylgenistin (mg)	11.67 ± 0.63
Glycitein (mg)	0.10 ± 0.01
Glycitin (mg)	0.77 ± 0.04
6″-O-Acetylglycitin (mg)	Trivial [3]
6″-O-Malonylglycitin (mg)	0.12 ± 0.01

[1]—Nutrient calculated from soy bread formulation using NDSR; [2]—Isoflavone content in soy breads (n = 6) determined using HPLC with PDA; [3]—Quantity less than level of quantification but detectable.

3.2. Soy Bread Palatability and Its Impact on Satiety

Sensory evaluation of soy bread using hedonic scoring indicated that the soy bread was palatable (Figure 2A). The radar analysis assessing bread attributes indicated that the taste of the soy bread was like their ideal bread, but the texture (hardness and chewiness) of the soy bread was significantly different ($p = 0.040$ and $p = 0.003$, respectively, Figure 2B) which is evidenced by a lower Hedonic score (likability) for texture being below neutral. Because of the 30% displacement of wheat flour by soy ingredients, CP participants detected a difference in the texture of the soy bread not being the same as their ideal bread and this was also reflected in the decrease in Hedonic score for texture. Loss in organoleptic qualities of wheat bread have been reported when 15% or 20% of the wheat flour is displaced by soy [40]. Moreover, there were no significant dose-dependent effects on bread palatability among the three cohorts nor the MTD participant. Therefore, consistent with our primary hypothesis, the sensory evaluation of the soy bread suggested that the bread would be well-tolerated and feasible for use in a short-term intervention study.

Visual analogue scale (VAS) was used to evaluate various aspects of satiety. The VAS scores indicate perceived thirst decreased significantly ($p < 0.001$) between one and three slices of bread after 8 h (Figure 2C). Dietary intake records show that dietary water and total beverage intake did not significantly decrease ($p = 0.065$ and $p = 0.152$, respectively) but there was a trend where mean dietary water intake decreased with increasing number of slices/day. Further, VAS scores indicate that there was a significant persistence in satiety with an increase in number of daily slices over time. Specifically, in the assessment of desire to eat, one slice/day compared to three slices/day at 8 h postprandial ($p < 0.001$) showed a decrease in the desire to eat (Figure 2B). Similar differences were observed with other satiety endpoints (Supplemental Figure S1A–G). The significant differences in perceived thirst and satiety suggest that soy bread may be contributing to fluid homeostasis in the gut. Although speculative, one possible rationale is the physicochemical differences between wheat and soy flour. Soy flour has three times the water-holding capacity than wheat flour and strong oil holding capacity [41,42]. This physicochemical property may facilitate water retention in the intestinal lumen, thereby enhancing satiety and fluid balance, but further dietary intervention studies using wheat control bread is warranted.

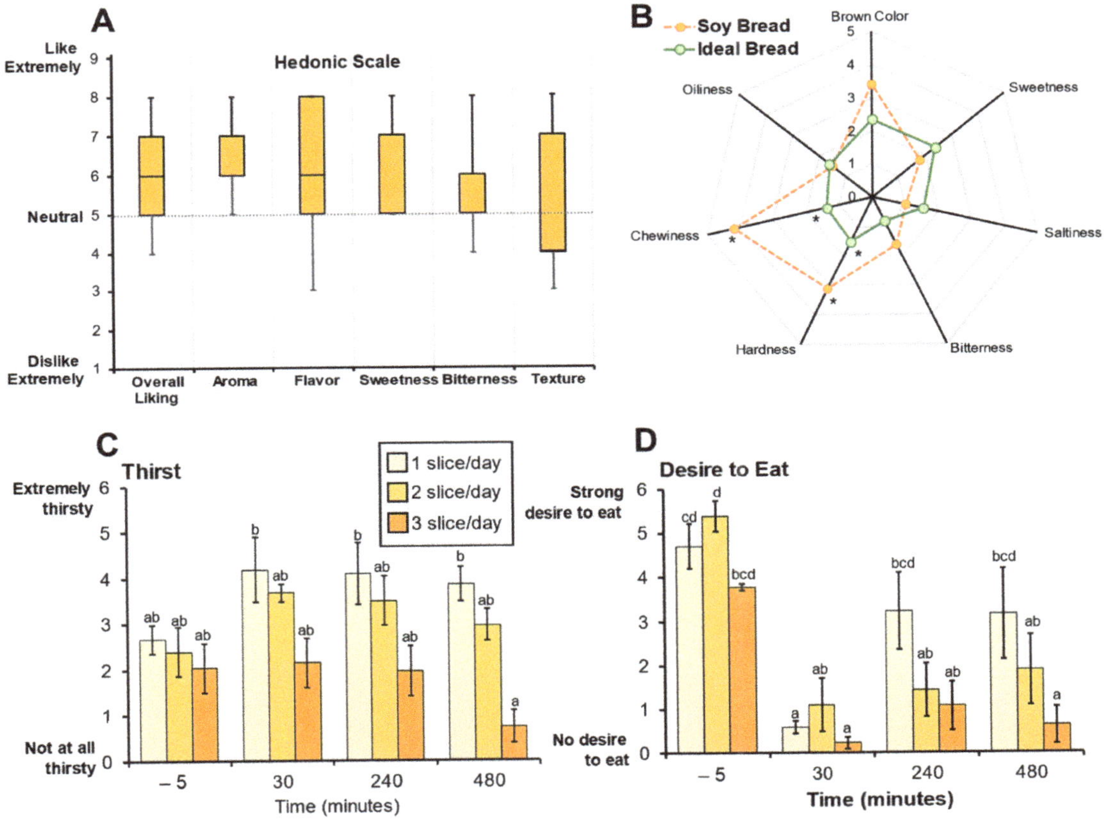

Figure 2. Sensory evaluations were used to assess soy bread palatability. (**A**) Hedonic scores representing soy bread palatability shown as box and whisker plots (median and interquartile ranges). (**B**) Radar analysis shows attribute intensities (median) of soy bread compared to participant's ideal bread. Visual Analogue Scale (VAS) scores for perceived thirst (**C**) and desire to eat (**D**) are represented as bar graphs (mean ± SD). Superscript letters represent the mean separation using ANOVA analysis and Tukey's post hoc test when significant differences ($p \leq 0.05$) were observed. * Significant differences ($p \leq 0.05$) between soy bread and ideal bread were assessed using a paired *t*-test.

3.3. Participant Clinical Features and Dietary Adhrence

Recruitment occurred over a 20-month period and a detailed CONSORT diagram was previously reported [36]. Of the 11 participants enrolled, 1 failed screening and withdrew before allocation, 9 completed the DE phase and 1 completed the MTD phase. Among the nine, two individuals from cohort 1 (1 slice/day) returned after 1 year to participate in cohort 3 (3 slices/day) (Figure 1). The age range was from 52 to 79 years old, and all were men except for one individual. A majority consumed their meals with lipase (Pancreatic Enzyme Replacement Therapy, PERT) supplementation (Table 2). Fasting blood chemistry, complete blood count, and renal and liver function tests were not significantly affected by soy bread intervention (Table 2). Although not a primary outcome in this small study, we observed a 18% reduction in mean fasting blood glucose in cohorts consuming two and three slices of soy bread per day. Previous animal studies report that isoflavones elicit hypoglycemic effects in rats with diabetes by inhibiting intestinal α-glucosidase activity [43] whereas others report that isoflavones improve insulin sensitivity by activating peroxisome proliferator activated receptors [44,45]. However, the anti-diabetic effects of soy isoflavones in human studies are mixed [46,47].

From three-day diet records, the total energy intake ($p = 0.095$) was not statistically significant with increases in soy bread dose, but this is likely attributed to the small size of this study. However, we observed a 9% and 35% reduction in mean energy intake in cohort 2 and 3 when compared to cohort 1. Likewise, there was a decrease, albeit not statistically significant, in daily water intake by ~35% in cohort 2 and 3 when compared to cohort 1. The dietary intake of total protein ($p = 0.013$), specifically animal protein ($p = 0.009$) decreased in a dose dependent manner (Table 2). Dietary intake was determined without the inclusion of soy bread. Participants were instructed to maintain their normal diets and consume breads in addition to their normal diets. Notably, the caloric content of the soy bread (weight for weight) is approximately 10% lower than a slice of conventional white or whole wheat bread. In a recent study, we demonstrated there is a wide range of caloric intake in subjects with CP, so this possible shift could have mixed consequences [48]. For those with substantial calories derived from lower quality food items using this bread product would be beneficial, whereas intake could be problematic in those with low caloric intake at baseline. Additionally, it will be necessary to assess tolerability over a long term, since it is anticipated that this (or any) intervention to effectively treat CP would need to last for years.

3.4. *Isoflavone Metabolism*

Oral ingestion of soy bread, when delivered in a dose escalating manner by increasing the number of bread slices per day, was associated with a linear increase (Pearson $R = 0.869$, $p = 0.0011$) in the excretion of total isoflavones in 24 h urine collections (Figure 3C). The total isoflavone excreted in each cohort during the DE phase was statistically significant, $p = 0.007$. The mean (±SEM) total daidzein excretion was 11.50 ± 3.55 mg/24 h for one slice, 19.02 ± 5.00 mg/24 h for two slices, and 30.51 ± 5.10 mg/24 h for three slices (Figure 3A), while total genistein was 4.66 ± 1.47 mg/24 h for one slice, 11.55 ± 3.05 mg/24 h for two slices, and 15.91 ± 2.95 mg/24 h for three slices (Figure 3B). A significant dose-dependent increase was observed in total daidzein ($p = 0.028$) excreted, but not for total genistein ($p = 0.188$). Daidzein metabolites such as dihydrodaidzein (DHD), O-desmethyl-angolensin (ODMA), and equol and genistein metabolites such as dihydrogenistein (DHG) and 6-OH-ODMA) were observed in 24 h urine samples.

3.5. *Isoflavonoid Phenotypes*

Phenotypes which favor daidzein excretion were predominant compared to genistein excreting phenotypes. Therefore, in this study population, 90% (9/10) had total daidzein: total genistein ratios (TD:TG ratio) greater than 1 (range 1.28–9.21). Like our previous soy bread studies [21,37], we observed four daidzein metabolizing phenotypes with this CP cohort. Specifically, among the ten 24 h urine samples analyzed, we observed one non-daidzein metabolizing (NDM) phenotype (subject 1G), three O-desmethyl-angolensin (ODMA: subjects 1D, 1E, and 2A), three ODMA + equol (subjects 1A, 1C, and 1H), and three equol (subjects 1B, 1F, and 1I) producing phenotypes (Figure 3A). The greatest excretion of total genistein was observed in NDM phenotype (subject 1G). Two subjects repeated 1 and 3 slice/day dosing arms. Both demonstrated that isoflavone metabolizing phenotypes are reproducible and metabolite excretion increased with dose. In the equol-producing subject

(1B and 1I) equol excretion increased from 28 (1 slice/day) to 61% (3 slice/day) of total isoflavones excreted, whereas in a ODMA producing subject (1A and 1H) their ODMA excretion increased from 3% (1 slice/day) to 10% (3 slice/day). In this same participant, equol was not present during their 1 slice/day period but detected during their 3 slice/day intervention period.

Table 2. Clinical and dietary information of chronic pancreatitis participants.

Participant Information	Dose Escalating Phase (Phase 1)			MTD [1] (Phase 2)
	1 Slice/Day (*n* = 3)	2 Slice/Day (*n* = 3)	3 Slice/Day (*n* = 3) [2]	3 Slice/Day (*n* = 1)
Clinical Features				
Body Mass Index (kg/m^2, mean ± SD)	23.4 ± 5.2	28.4 ± 1.6	26.5 ± 3.9	24.6
Lipase (PERT) with meals (%)	66	100	66	yes
Lipase dose range (1000 units/meal)	24 and 72	24 to 48	24 and 72	24
Diabetes mellitus (%)	33	100	66	no
Pancreatic calcification/calculi (%)	100	100	100	yes
Main Pancreatic Duct Dilatation (%)	66	66	100	no
Clinical Laboratory Values (mean ± SD) [3]				
Creatinine (mg/dL)				
Pre-intervention	1.03 ± 0.06	1.17 ± 0.02	0.94 ± 0.09	1.49
Post-intervention	0.09 ± 0.06	1.24 ± 0.05	0.89 ± 0.07	1.49
Blood glucose (mg/dL)				
Pre-intervention	104 ± 6	172 ± 52	236 ± 141	82
Post-intervention	113 ± 8	139 ± 24	194 ± 79	89
Alanine aminotransferase (IU/L)				
Pre-intervention	16 ± 5	22 ± 10	17 ± 2	12
Post-intervention	18 ± 4	26 ± 10	17 ± 3	16
Aspartate aminotransferase (IU/L)				
Pre-intervention	16 ± 1	18 ± 4	17 ± 1	19
Post-intervention	22 ± 4	19 ± 4	18 ± 3	21
Total bilirubin (mg/dL)				
Pre-intervention	0.6 ± 0.1	0.4 ± 0.1	0.7 ± 0.1	0.4
Post-intervention	0.5 ± 0.1	0.4 ± 0.1	0.7 ± 0.1	0.4
Albumin (g/dL)				
Pre-intervention	4.3 ± 0.2	4.3 ± 0.3	4.1 ± 0.2	3.6
Post-intervention	4.3 ± 0.2	4.3 ± 0.3	4.1 ± 0.2	3.6
Alkaline phosphatase (IU/L)				
Pre-intervention	74 ± 12	110 ± 25	101 ± 45	98
Post-intervention	70 ± 7	109 ± 16	108 ± 34	101
Daily Nutrient Intake (mean ± SEM) [4]				
Total Energy (kcal)	2300 ± 288	2095 ± 240	1517 ± 221	1592 ± 158
Total fat (g)	95.0 ± 16.8	75.2 ± 11.7	65.7 ± 14.4	49.0 ± 18.6
Cholesterol (mg)	391.4 ± 109.5	174.3 ± 33.2	216.9 ± 64.5	81.6 ± 71.1
Total carbohydrate (g)	278.1 ± 32.2	288.8 ± 38.5	191.5 ± 31.8	252.7 ± 46.3
Dietary fiber (g)	21.4 ± 5.6	18.4 ± 2.9	25.6 ± 5.5	24.9 ± 11.3
Total protein(g) [5]	96.7 ± 13.4 ^{a}	74.6 ± 8.3 ^{ab}	48.2 ± 9.6 ^{b}	50.1 ± 12.1
Animal protein (g) [5]	74.7 ± 11.3 ^{a}	46.4 ± 9.6 ^{ab}	28.6 ± 7.7 ^{b}	18.5 ± 12.6
Vegetable protein (g)	22.0 ± 3.6	28.2 ± 3.3	19.6 ± 3.4	29.1 ± 8.2
Total water (g)	3296 ± 936	2149 ± 517	2083 ± 514	2143 ± 189

[1]—Maximum Tolerated Dose; [2]—2 of 3 subjects were also in the 1 slice/day subgroup; [3]—Fasting blood serum; [4]—Data collected from 3-day diet records, not including intake of soy bread, and analyzed using NDSR (Nutrition Data System for Research, Minnesota, MN, USA); [5]—Statistically significant ($p < 0.05$) ANOVA analysis and mean separation are indicated by different superscript letters.

Isoflavone absorption and metabolism patterns in our cohort were similar with those consuming a habitual Asian diet [49] and those consuming a Westernized diet after soy intervention [21,37]. Repeated daily dosing in both 1-week and 4-week interventions demonstrated both time periods were sufficient in duration to capture isoflavone metabolites in urine. Isoflavone microbial metabolites have shown greater bioactivity than their parent compounds (daidzein, genistein, and glycitein) [50] and their production shows great interspecies and regional diversity. Among mammalian species, the frequency of equol production is much higher in rodents than in humans, more frequent in Asian than Western populations, and much higher in vegetarian diets [49,51]. In our cohort, equol-producing compared to NDM and ODMA phenotypes had the greatest daily intake of total dietary fiber ($p = 0.001$) and insoluble fiber ($p = 0.002$). The NDM (subject 1G) phenotype had the lowest caloric intake at ~800 kcal/day whereas the mean daily energy intake for the other phenotypes was ~2000 kcal/day.

3.6. Potential Implications for Clinical Care

CP is a debilitating, irreversible disease with a detrimental impact on quality of life, nutrition, and health care visits. While there are currently no approved medical therapies to interrupt this chronic inflammatory process, much of the nutritional management of chronic pancreatitis is to prevent malnutrition by treating the symptoms of maldigestion and malabsorption using medical modalities and diet modification (high protein, high energy diet) [14,52]. However, this fails to address a number of the underlying pathways that contribute to disease progression and clinical manifestations, including inflammatory pathways and gut dysbiosis, amongst others. Advances in microbiome research and whole genome sequencing have provided insights into other alternative approaches, such as restoration of intestinal functional capacity and gut health to improve the nutritional status in CP, and warrant further investigation in CP [53–55]. Among the many food ingredients, plant-based foods offer a unique opportunity to enhance bioactive profiles in the food ingredients through horticultural conditions, biofortification, and precision breeding [56–59]. Moreover, complementary food processing techniques can enhance bioavailability of bioactive compounds or utilize food structures to modulate their delivery to localized sites along the gastrointestinal tract [38,60]. While nutritional intervention studies represent a great opportunity in CP, there are several challenges, including the lack of standardized outcome assessments to determine clinical efficacy [61]. Nevertheless, considering the absence of approved medical therapies for CP and the significant nutritional abnormalities, this is a key opportunity for future investigations.

The current pilot trial was primarily focused on assessing feasibility, safety, and tolerability of the soy-enriched bread in a study population with chronic pancreatitis. Due to the small sample size and expected within-group variability, we were unable to characterize the mechanisms of cellular action or immunological signaling, and examined isoflavone metabolism as a biochemical outcome. Nevertheless, a prior pre-clinical resource using a similar diet in an animal model demonstrated that dietary soy modulates natural killer cell function, including reduced expression of interferon-gamma, a key mediator for chronic pancreatitis [62].

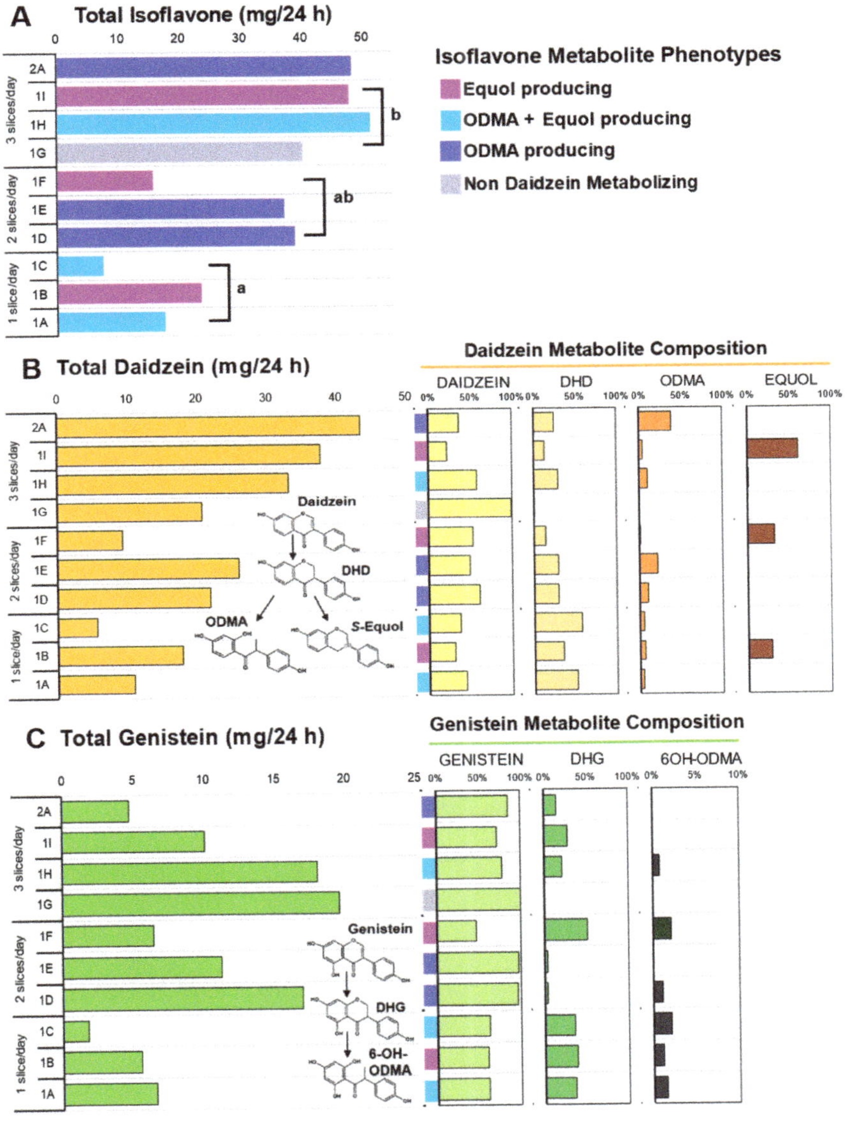

Figure 3. Isoflavone metabolites in 24 h urine collection across three doses of soy bread. (**A**) Total isoflavones excreted across three doses of soy bread and colors represent the four daidzein excreting phenotypes observed in this cohort. (**B**) Total daidzein concentration excreted (**left**) and their corresponding metabolite composition of each subject expressed as a percentage of total daidzein (**right**). (**C**) Total genistein concentration excreted (**left**) and their corresponding metabolite composition of each subject expressed as a percentage of total genistein (**right**). Superscript letters represent the mean separation using ANOVA analysis and Tukey's posthoc test when significant differences ($p \leq 0.05$) were observed. Isoflavone structures show abbreviated metabolite pathways (inset). DHD: dihydrodaidzein; DHG: dihydrogenistein; ODMA: O-desmethyl-angolensin; 6-OH-ODMA: 6-hydroxyl ODMA.

4. Conclusions

In this small pilot study, short-term soy intervention with soy bread in participants with CP demonstrated that the soy bread effectively delivered isoflavones in a dose-dependent manner. To our knowledge, this study is the first to report isoflavone metabolism and isoflavone phenotypes in patients with chronic pancreatitis. Because of the great heterogeneity in CP disease, metabolic phenotypes of soy isoflavone intervention are critical to help decipher heterogeneity in biologic responses among individuals with CP. This study combined with our previous findings demonstrating safety and the anti-inflammatory effects of the soy bread provides the necessary foundation to design future foods for CP and larger clinical trials.

Supplementary Materials: The following supporting information can be downloaded at: https://www.mdpi.com/article/10.3390/foods12091762/s1. Figure S1: Supplementary VAS Satiety Assessment.

Author Contributions: Conceptualization (J.H.A.-J., E.L., G.B.L., D.L.C., P.A.H. and Y.V.); Data curation (J.H.A.-J., E.L., D.S. and P.A.H.); Formal analysis (all authors); Funding acquisition (G.B.L., D.L.C., P.A.H. and Y.V.); Investigation (J.H.A.-J., E.L. and P.A.H.); Project administration (J.H.A.-J., E.L., D.L.C. and P.A.H.); Resources (all authors); Supervision (J.H.A.-J., E.L., G.B.L., D.L.C., P.A.H. and Y.V.); Validation (J.H.A.-J., E.L., G.B.L., D.L.C., P.A.H. and Y.V.); Visualization (J.H.A.-J.); Writing—original draft (J.H.A.-J., P.A.H., D.S. and Y.V.); Writing—review and editing (all authors). All authors have read and agreed to the published version of the manuscript.

Funding: This study was funded by a pilot grant from the American College of Gastroenterology. Research reported in this publication was also supported by the National Cancer Institute and the National Institute of Diabetes and Digestive and Kidney Diseases (NIDDK) under award number U01DK108327 (D.C. and P.H.). The project described was also supported by Award Number Grant UL1TR002733 from the National Center for Advancing Translational Sciences and the Center for Advanced Functional Foods Research and Entrepreneurship (CAFFRE). The content is solely the responsibility of the authors and does not necessarily represent the official views of the National Institutes of Health.

Institutional Review Board Statement: The study was conducted in accordance with the Declaration of Helsinki, and approved by The Ohio State University Biomedical Science Institutional Review Board (approved 1 October 2014) as protocol 2014H0226.

Data Availability Statement: All related data and methods are presented in this paper. Additional inquiries should be addressed to the corresponding author.

Acknowledgments: The authors are grateful to the participants and study coordinator team from the Division of Gastroenterology, Hepatology, and Nutrition who made this study possible.

Conflicts of Interest: The authors declare no conflict of interest.

Abbreviations

DHD: dihydrodaidzein; DHG, dihydrogenistein; HPLC, High Performance Liquid Chromatography; NDM, non-daidzein metabolizing; ODMA, O-desmethylangolensin; 6-OH-ODMA, 6-hydroxy-O-desmethylangolensin; PDA, Photodiode array; PERT, Pancreatic Enzyme Replacement Therapy; VAS, Visual Analogue Scale.

References

1. Kleeff, J.; Whitcomb, D.C.; Shimosegawa, T.; Esposito, I.; Lerch, M.M.; Gress, T.; Mayerle, J.; Drewes, A.M.; Rebours, V.; Akisik, F.; et al. Chronic pancreatitis. *Nat. Rev. Dis. Primers* **2017**, *3*, 17060. [CrossRef] [PubMed]
2. Hart, P.A.; Conwell, D.L. Chronic Pancreatitis: Managing a Difficult Disease. *Am. J. Gastroenterol.* **2020**, *115*, 49–55. [CrossRef] [PubMed]
3. Rasmussen, H.H.; Irtun, O.; Olesen, S.S.; Drewes, A.M.; Holst, M. Nutrition in chronic pancreatitis. *World J. Gastroenterol.* **2013**, *19*, 7267–7275. [CrossRef] [PubMed]
4. Duggan, S.N. Negotiating the complexities of exocrine and endocrine dysfunction in chronic pancreatitis. *Proc. Nutr. Soc.* **2017**, *76*, 484–494. [CrossRef] [PubMed]

5. Ramsey, M.L.; Conwell, D.L.; Hart, P.A. Complications of chronic pancreatitis. *Dig. Dis. Sci.* **2017**, *62*, 1745–1750. [CrossRef]
6. Capurso, G.; Signoretti, M.; Archibugi, L.; Stigliano, S.; Fave, G.D. Systematic review and meta-analysis: Small intestinal bacterial overgrowth in chronic. Pancreatitis. *United Eur. Gastroenterol. J.* **2016**, *4*, 697–705. [CrossRef]
7. Jandhyala, S.M.; Madhulika, A.; Deepika, G.; Rao, G.V.; Reddy, D.N.; Subramanyam, C.; Sasikala, M.; Talukdar, R. Altered intestinal microbiota in patients with chronic pancreatitis: Implications in diabetes and metabolic abnormalities. *Sci. Rep.* **2017**, *7*, srep43640. [CrossRef]
8. Schepis, T.; De Lucia, S.S.; Nista, E.C.; Manilla, V.; Pignataro, G.; Ojetti, V.; Piccioni, A.; Gasbarrini, A.; Franceschi, F.; Candelli, M. Microbiota in pancreatic diseases: A review of the literature. *J. Clin. Med.* **2021**, *10*, 5920. [CrossRef]
9. Komar, H.M.; Hart, P.A.; Cruz-Monserrate, Z.; Conwell, D.L.; Lesinski, G.B. Local and systemic expression of immunomodulatory factors in chronic pancreatitis. *Pancreas* **2017**, *46*, 986. [CrossRef]
10. Frost, F.; Weiss, F.U.; Sendler, M.; Kacprowski, T.; Rühlemann, M.; Bang, C.; Franke, A.; Völker, U.; Völzke, H.; Lamprecht, G.; et al. The gut microbiome in patients with chronic pancreatitis is characterized by significant dysbiosis and overgrowth by opportunistic pathogens. *Clin. Transl. Gastroenterol.* **2020**, *11*, e00232. [CrossRef]
11. Petrov, M.S. Metabolic trifecta after pancreatitis: Exocrine pancreatic dysfunction, altered gut microbiota, and new-onset diabetes. *Clin. Transl. Gastroenterol.* **2019**, *10*, e00086. [CrossRef]
12. De Kort, S.; Keszthelyi, D.; Masclee, A.A.M. Leaky gut and diabetes mellitus: What is the link? *Obes. Rev.* **2011**, *12*, 449–458. [CrossRef]
13. Canani, R.B.; Di Costanzo, M.; Leone, L.; Pedata, M.; Meli, R.; Calignano, A. Potential beneficial effects of butyrate in intestinal and extraintestinal diseases. *World J. Gastroenterol.* **2011**, *17*, 1519–1528. [CrossRef]
14. Arvanitakis, M.; Ockenga, J.; Bezmarevic, M.; Gianotti, L.; Krznarić, Ž.; Lobo, D.N.; Löser, C.; Madl, C.; Meier, R.; Phillips, M.; et al. ESPEN guideline on clinical nutrition in acute and chronic pancreatitis. *Clin. Nutr. Open Sci.* **2020**, *39*, 612–631. [CrossRef]
15. Eldridge, A.C.; Black, L.T.; Wolf, W.J. Carbohydrate composition of soybean flours, protein concentrates, and isolates. *J. Agric. Food Chem.* **1979**, *27*, 799–802. [CrossRef]
16. Redondo-Cuenca, A.; Villanueva-Suárez, M.J.; Rodríguez-Sevilla, M.D.; Mateos-Aparicio, I. Chemical composition and dietary fibre of yellow and green commercial soybeans (*Glycine max*). *Food Chem.* **2007**, *101*, 1216–1222. [CrossRef]
17. Kocot, A.M.; Jarocka-Cyrta, E.; Drabińska, N. Overview of the importance of biotics in gut barrier integrity. *Int. J. Mol. Sci.* **2022**, *23*, 2896. [CrossRef]
18. Sheflin, A.M.; Melby, C.L.; Carbonero, F.; Weir, T.L. Linking dietary patterns with gut microbial composition and function. *Gut Microbes* **2017**, *8*, 113–129. [CrossRef]
19. Wu, Q.; Chen, T.; El-Nezami, H.; Savidge, T.C. Food ingredients in human health: Ecological and metabolic perspectives implicating gut microbiota function. *Trends Food Sci. Technol.* **2020**, *100*, 103–117. [CrossRef]
20. Clavel, T.; Fallani, M.; Lepage, P.; Levenez, F.; Mathey, J.; Rochet, V.; Serezat, M.; Sutren, M.; Henderson, G.; Bennetau-Pelissero, C.; et al. Isoflavones and functional foods alter the dominant intestinal microbiota in postmenopausal women. *J. Nutr.* **2005**, *135*, 2786–2792. [CrossRef]
21. Ahn-Jarvis, J.; Clinton, S.K.; Riedl, K.M.; Vodovotz, Y.; Schwartz, S.J. Impact of food matrix on isoflavone metabolism and cardiovascular biomarkers in adults with hypercholesterolemia. *Food Funct.* **2012**, *3*, 1051–1058. [CrossRef] [PubMed]
22. Marciani, L.; Pritchard, S.E.; Hellier-Woods, C.; Costigan, C.; Hoad, C.L.; Gowland, P.A.; Spiller, R.C. Delayed gastric emptying and reduced postprandial small bowel water content of equicaloric whole meal bread versus rice meals in healthy subjects: Novel MRI insights. *Eur. J. Clin. Nutr.* **2013**, *67*, 754–758. [CrossRef] [PubMed]
23. Setchell, K.D.; Brown, N.M.; Lydeking-Olsen, E. The clinical importance of the metabolite equol—A clue to the effectiveness of soy and its isoflavones. *J. Nutr.* **2002**, *132*, 3577–3584. [CrossRef] [PubMed]
24. Sánchez-Calvo, J.M.; Rodríguez-Iglesias, M.A.; Molinillo, J.M.; Macías, F.A. Soy isoflavones and their relationship with microflora: Beneficial effects on human health in equol producers. *Phytochem. Rev.* **2013**, *12*, 979–1000. [CrossRef]
25. Mayo, B.; Vázquez, L.; Flórez, A.B. Equol: A bacterial metabolite from the daidzein isoflavone and its presumed beneficial health effects. *Nutrients* **2019**, *11*, 2231. [CrossRef]
26. Rowland, I.R.; Wiseman, H.; Sanders, T.A.; Adlercreutz, H.; Bowey, E.A. Interindividual variation in metabolism of soy isoflavones and lignans: Influence of habitual diet on equol production by the gut microflora. *Nutr. Cancer* **2000**, *36*, 27–32. [CrossRef]
27. Akaza, H.; Miyanaga, N.; Takashima, N.; Naito, S.; Hirao, Y.; Tsukamoto, T.; Fujioka, T.; Mori, M.; Kim, W.J.; Song, J.M.; et al. Comparisons of percent equol producers between prostate cancer patients and controls: Case-controlled studies of isoflavones in Japanese, Korean and American residents. *Jpn. J. Clin. Oncol.* **2004**, *34*, 86–89. [CrossRef]
28. Iino, C.; Shimoyama, T.; Iino, K.; Yokoyama, Y.; Chinda, D.; Sakuraba, H.; Fukuda, S.; Nakaji, S. Daidzein intake is associated with equol producing status through an increase in the intestinal bacteria responsible for equol production. *Nutrients* **2019**, *11*, 433. [CrossRef]
29. Yoshikata, R.; Myint, K.Z.; Ohta, H.; Ishigaki, Y. Inter-relationship between diet, lifestyle habits, gut microflora, and the equol-producer phenotype: Baseline findings from a placebo-controlled intervention trial. *Menopause* **2019**, *26*, 273–285. [CrossRef]
30. Wells, C.L.; Jechorek, R.P.; Kinneberg, K.M.; Debol, S.M.; Erlandsen, S.L. The isoflavone genistein inhibits internalization of enteric bacteria by cultured Caco-2 and HT-29 enterocytes. *J. Nutr.* **1999**, *129*, 634–640. [CrossRef]
31. Abron, J.D.; Singh, N.P.; Price, R.L.; Nagarkatti, M.; Nagarkatti, P.S.; Singh, U.P. Genistein induces macrophage polarization and systemic cytokine to ameliorate experimental colitis. *PLoS ONE* **2018**, *13*, e0199631. [CrossRef]

32. Sato, Y.; Itagaki, S.; Oikawa, S.; Ogura, J.; Kobayashi, M.; Hirano, T.; Sugawara, M.; Iseki, K. Protective effect of soy isoflavone genistein on ischemia-reperfusion in the rat small intestine. *Biol. Pharm. Bull.* **2011**, *34*, 1448–1454. [CrossRef]
33. Wu, Z.Y.; Sang, L.X.; Chang, B. Isoflavones and inflammatory bowel disease. *World J. Clin. Cases* **2020**, *8*, 2081–2091. [CrossRef]
34. Jalili, M.; Vahedi, H.; Janani, L.; Poustchi, H.; Malekzadeh, R.; Hekmatdoost, A. Soy isoflavones supplementation for patients with irritable bowel syndrome: A randomized double blind clinical trial. *Middle East J. Dig. Dis.* **2015**, *7*, 170–176.
35. Jalili, M.; Vahedi, H.; Poustchi, H.; Hekmatdoost, A. Soy isoflavones and cholecalciferol reduce inflammation, and gut permeability, without any effect on antioxidant capacity in irritable bowel syndrome: A randomized clinical trial. *Clin. Nutr. ESPEN* **2019**, *34*, 50–54. [CrossRef]
36. Ahn-Jarvis, J.; Lombardo, E.; Cruz-Monserrate, Z.; Badi, N.; Crowe, O.; Kaul, S.; Komar, H.; Krishna, S.G.; Lesinski, G.B.; Mace, T.A.; et al. Reduction of inflammation in chronic pancreatitis using a soy bread intervention: A feasibility study. *Pancreatology* **2020**, *20*, 852–859. [CrossRef]
37. Ahn-Jarvis, J.H.; Clinton, S.K.; Grainger, E.M.; Riedl, K.M.; Schwartz, S.J.; Lee, M.L.T.; Cruz-Cano, R.; Young, G.S.; Lesinski, G.B.; Vodovotz, Y. Isoflavone Pharmacokinetics and Metabolism after Consumption of a Standardized Soy and Soy–Almond Bread in Men with Asymptomatic Prostate Cancer. *Cancer Prev. Res.* **2015**, *8*, 1045–1054. [CrossRef]
38. Ahn-Jarvis, J.H.; Riedl, K.M.; Schwartz, S.J.; Vodovotz, Y. Design and selection of soy breads used for evaluating isoflavone bioavailability in clinical trials. *J. Agric. Food Chem.* **2013**, *61*, 3111–3120. [CrossRef]
39. Grainger, E.M.; Moran, N.E.; Francis, D.M.; Schwartz, S.J.; Wan, L.; Thomas-Ahner, J.; Kopec, R.E.; Riedl, K.M.; Young, G.S.; Abaza, R.; et al. A novel tomato-soy juice induces a dose-response increase in urinary and plasma phytochemical biomarkers in men with prostate cancer. *J. Nutr.* **2019**, *149*, 26–35. [CrossRef]
40. Dhingra, S.; Jood, S. Organoleptic and nutritional evaluation of wheat breads supplemented with soybean and barley flour. *Food Chem.* **2002**, *77*, 479–488. [CrossRef]
41. Ahn-Jarvis, J.H.; Schwartz, S.J.; Vodovotz, Y. Optimizing isoflavone-rich food delivery systems for human clinical trials. In *Isoflavones: Chemistry, Analysis, Function and Effects*; Preedy, V., Ed.; Royal Society of Chemistry Publishing: Cambridge, UK, 2012; pp. 399–422. [CrossRef]
42. Ratnawati, L.; Desnilasari, D.; Surahman, D.N.; Kumalasari, R. Evaluation of physicochemical, functional and pasting properties of soybean, mung bean and red kidney bean flour as ingredient in biscuit. *IOP Conf. Ser. Earth Environ. Sci.* **2019**, *251*, 012026. [CrossRef]
43. Jin, M.; Shen, M.H.; Jin, M.H.; Jin, A.H.; Yin, X.Z.; Quan, J.S. Hypoglycemic property of soy isoflavones from hypocotyl in Goto-Kakizaki diabetic rats. *J. Clin. Biochem. Nutr.* **2018**, *62*, 148–154. [CrossRef] [PubMed]
44. Ronis, M.J.; Chen, Y.; Badeaux, J.; Badger, T.M. Dietary soy protein isolate attenuates metabolic syndrome in rats via effects on PPAR, LXR, and SREBP signaling. *J. Nutr.* **2009**, *139*, 1431–1438. [CrossRef] [PubMed]
45. Mezei, O.; Banz, W.J.; Steger, R.W.; Peluso, M.R.; Winters, T.A.; Shay, N. Soy isoflavones exert antidiabetic and hypolipidemic effects through the PPAR pathways in obese Zucker rats and murine RAW 264.7 cells. *J. Nutr.* **2003**, *133*, 1238–1243. [CrossRef] [PubMed]
46. Jayagopal, V.; Albertazzi, P.; Kilpatrick, E.S.; Howarth, E.M.; Jennings, P.E.; Hepburn, D.A.; Atkin, S.L. Beneficial effects of soy phytoestrogen intake in postmenopausal women with type 2 diabetes. *Diabetes Care* **2002**, *25*, 1709–1714. [CrossRef] [PubMed]
47. Liu, Z.M.; Chen, Y.M.; Ho, S.C. Effects of soy intake on glycemic control: A meta-analysis of randomized controlled trials. *Am. J. Clin. Nutr.* **2011**, *93*, 1092–1101. [CrossRef]
48. Roberts, K.M.; Golian, P.; Nahikian-Nelms, M.; Hinton, A.; Madril, P.; Basch, K.; Conwell, D.; Hart, P.A. Does the Healthy Eating Index and Mediterranean Diet Score identify the nutritional adequacy of dietary patterns in chronic pancreatitis? *Dig. Dis. Sci.* **2019**, *64*, 2318–2326. [CrossRef]
49. Adlercreutz, H.; Honjo, H.; Higashi, A.; Fotsis, T.; Hämäläinen, E.; Hasegawa, T.; Okada, H. Urinary excretion of lignans and isoflavonoid phytoestrogens in Japanese men and women consuming a traditional Japanese diet. *Am. J. Clin. Nutr.* **1991**, *54*, 1093–1100. [CrossRef]
50. Setchell, K.D.; Clerici, C. Equol: History, chemistry, and formation. *J. Nutr.* **2010**, *140*, 1355S–1362S. [CrossRef]
51. Gu, L.; House, S.E.; Prior, R.L.; Fang, N.; Ronis, M.J.; Clarkson, T.B.; Wilson, M.E.; Badger, T.M. Metabolic phenotype of isoflavones differ among female rats, pigs, monkeys, and women. *J. Nutr.* **2006**, *136*, 1215–1221. [CrossRef]
52. Roberts, K.M.; Hart, P.A.; Duggan, S. Revisiting Dietary Approaches in the Management of Chronic Pancreatitis. *Curr. Treat. Options Gastroenterol.* **2022**, *20*, 605–623. [CrossRef]
53. Holscher, H.D.; Caporaso, J.G.; Hooda, S.; Brulc, J.M.; Fahey Jr, G.C.; Swanson, K.S. Fiber supplementation influences phylogenetic structure and functional capacity of the human intestinal microbiome: Follow-up of a randomized controlled trial. *Am. J. Clin. Nutr.* **2015**, *101*, 55–64. [CrossRef]
54. Lynch, S.V.; Pedersen, O. The human intestinal microbiome in health and disease. *N. Engl. J. Med.* **2016**, *375*, 2369–2379. [CrossRef]
55. Salminen, S.; Bouley, C.; Boutron, M.C.; Cummings, J.H.; Franck, A.; Gibson, G.R.; Isolauri, E.; Moreau, M.C.; Roberfroid, M.; Rowland, I. Functional food science and gastrointestinal physiology and function. *Br. J. Nutr.* **1998**, *80*, S147–S171. [CrossRef]
56. Bulgari, R.; Franzoni, G.; Ferrante, A. Biostimulants application in horticultural crops under abiotic stress conditions. *Agronomy* **2019**, *9*, 306. [CrossRef]

57. Newman, R.; Waterland, N.; Moon, Y.; Tou, J.C. Selenium biofortification of agricultural crops and effects on plant nutrients and bioactive compounds important for human health and disease prevention–a review. *Plant Foods Hum. Nutr.* **2019**, *74*, 449–460. [CrossRef]
58. Kumar, S.; Pandey, G. Biofortification of pulses and legumes to enhance nutrition. *Heliyon* **2020**, *6*, e03682. [CrossRef]
59. Tuncel, A.; Corbin, K.R.; Ahn-Jarvis, J.; Harris, S.; Hawkins, E.; Smedley, M.A.; Harwood, W.; Warren, F.J.; Patron, N.J.; Smith, A.M. Cas9-mediated mutagenesis of potato starch-branching enzymes generates a range of tuber starch phenotypes. *Plant Biotechnol. J.* **2019**, *17*, 2259–2271. [CrossRef]
60. Parada, J.; Santos, J.L. Interactions between starch, lipids, and proteins in foods: Microstructure control for glycemic response modulation. *Crit. Rev. Food Sci. Nutr.* **2016**, *56*, 2362–2369. [CrossRef]
61. Hart, P.A.; Andersen, D.K.; Lyons, E.; Cote, G.A.; Cruz-Monserrate, Z.; Dworkin, R.H.; Elmunzer, B.J.; Fogel, E.L.; Forsmark, C.E.; Gilron, I.; et al. Clinical Trials in Pancreatitis: Opportunities and Challenges in the Design and Conduct of Patient-Focused Clinical Trials in Recurrent Acute and Chronic Pancreatitis: Summary of a National Institute of Diabetes and Digestive and Kidney Diseases Workshop. *Pancreas* **2022**, *51*, 715–722. [CrossRef]
62. Mace, T.A.; Ware, M.B.; King, S.A.; Loftus, S.; Farren, M.R.; McMichael, E.; Scoville, S.; Geraghty, C.; Young, G.; Carson III, W.E.; et al. Soy isoflavones and their metabolites modulate cytokine-induced natural killer cell function. *Sci. Rep.* **2019**, *9*, 5068. [CrossRef] [PubMed]

MDPI
St. Alban-Anlage 66
4052 Basel
Switzerland
www.mdpi.com

Foods Editorial Office
E-mail: foods@mdpi.com
www.mdpi.com/journal/foods

www.ingramcontent.com/pod-product-compliance
Lightning Source LLC
Chambersburg PA
CBHW062033220926
51850CB00010B/2527
9783036599052